DATE DUE

AP 30 '97			

DEMCO 38-296

The Institute of British Geographers
Studies in Geography

The Changing Geography of China

IBG STUDIES IN GEOGRAPHY

General Editors
Felix Driver and Neil Roberts

IBG Studies in Geography are a range of stimulating texts which critically summarize the latest developments across the entire field of geography. Intended for students around the world, the series is published by Blackwell Publishers on behalf of the Institute of British Geographers.

Published

Debt and Development
Stuart Corbridge
The Changing Geography of China
Frank Leeming
Critical Issues in Tourism
Gareth Shaw and Allan M. Williams
The European Community
Allan M. Williams

In preparation

Geography and Gender
Liz Bondi
Population Geography
A. G. Champion, A. Findlay and E. Graham
Rural Geography
Paul Cloke
The Geography of Crime and Policing
Nick Fyfe
Fluvial Geomorphology
Keith Richards
Russia in the Modern World
Denis Shaw
A Geography of Housing
Susan Smith
The Sources and Uses of Energy
John Soussan
Retail Restructuring
Neil Wrigley

THE CHANGING GEOGRAPHY OF CHINA

Frank Leeming

Blackwell Publishers
108 Cowley Road
Oxford OX4 1JF
UK

238 Main Street
Cambridge, Massachusetts 02142
USA

First published 1993

Reprinted with corrections
and Guide to Further Reading 1993

British Library Cataloguing in Publication Data

A CIP catalogue record for this book is available from the British Library.

Library of Congress Cataloging-in-Publication Data

Leeming, Frank.
    The changing geography of China / Frank Leeming.
    p. cm.—(IBG studies in geography)
    Includes bibliographical references and index.
    ISBN 0–631–17675–6 (hb).—ISBN 0–631–18137–7 (pb)
    1. China—Geography. I. Title. II. Series.
    DS706.7.L44 1993
    915.1—dc20   92-14969 CIP

This book is printed on acid-free paper

Typeset in 11 on 13 pt Plantin
by TecSet Ltd, Wallington, Surrey

Printed in Great Britain by T. J. Press Ltd, Padstow, Cornwall

Contents

Plates

Figures

Tables

ONE

China of Many Faces

China's Status in Eurasia

China is one of humanity's principal homelands on the planet Earth, sharing that status with India, the Middle East and Europe. All four of these peripheries of the Eurasian continent have supported varied and persistent civilizations maintained by large populations; all have contributed richly to the common inheritance of mankind during the past three or more thousand years; and all have contributed to the construction of civilization in various other parts of the world, whether in the Americas, Japan, south-east Asia or Africa. Of these four homelands, China is not the oldest established (that distinction belongs to the Middle East), but it is the most individual and has the most continuous history since Neolithic times and even before. All the evidence is that in northern China at least, human populations have always been of Chinese physical type and have always used languages akin to Chinese. In this respect traditional history and modern archaeology are agreed. In addition, of all the great homelands of humanity, China is the one which at the present time supports the largest human population – around 1160 million people, about one-fifth of the whole human family. Remarkably, there has probably never been a time when China's human population fell much, or for long, below that proportion of all humanity.

Until modern times, China's development was exceptionally free of outside influences. This was one outcome of the location of the Chinese subcontinent. Located in east Asia, facing the Pacific and reaching inland to the vast plateaus, basins and deserts of central Asia, China had few neighbours, and until the modernization of Japan in the late nineteenth century, none of comparable status. South-east Asia particularly, apart from Java, remained generally scantily peopled until the nineteenth century. Of all the four great peripheral homelands in this Eurasian fringe, China was the one (except Japan) with fewest neighbours and the one most isolated from currents of communication. In traditional times Japan, isolated from everywhere except China, owed practically everything to the adoption of Chinese systems. In location terms, China was a terminus.

China occupies around 7 per cent of the world's land area, the largest national space after Russia and Canada, and about the same proportion as the United States. A subcontinent as vast as the Chinese cannot but be immensely varied. China stretches through some 36 degrees of latitude from north to south, and around 60 degrees of longitude from east to west – comparable in terms of latitude with London almost to Senegal or from Labrador to Jamaica; in terms of longitude from San Francisco to Boston, or from London to the Urals. In China, natural conditions range from subtropical jungles in the far south to permafrost on the Soviet frontier in the north-east, and rock or sand desert in the western interior. Nor are these dramatic extremes the only phases of regional variety; a Chinese province is typically of the same order of size as a European country, and most provinces display diversities of natural environment and human management as great as those in European countries such as Britain or Italy. In Hubei province for example, on the Yangzi in central China, which has around 53 million people, several regions are distinguished – the watery Yangzi landscape with many lakes around Wuhan; the formidable mountains of the west, including the famous Yangzi gorges; the broad valley of the Han river to the north-west, and the hill country on the north-east and south-east borders of the province. Within each of these regions individual counties have their own geographies and histories: Hubei has 73 counties. These Chinese counties are typically similar in size, though not necessarily in population, to British or United States counties.

China as a Homeland of Humanity

China, with 1143 million people in 1990, has the largest national population on earth by far. Among human populations as among those of other kinds of animal, large population totals and heavy densities are evidence of highly successful biological relationships with environments, together with successful systems of social organization, so far without important positive checks to growth. How can this idea be squared with Chinese poverty?

China's modern population growth has taken place over several centuries of time, in a phase in which, due to changes in the Western and international worlds, perceptions of wealth and poverty were in rapid flux. By 1750, when the Western industrial adventure which shaped the modern world was in its earliest stages, China's population had already reached the exceptional size of around 250 million. That China's population rose to a figure of this kind so early in modern times did arise from exceptionally fruitful relations between community and environment, together with very successful systems of social organization. China in traditional times had labour-intensive schemes of agriculture and industry which for many generations continued to expand job opportunities; family-oriented social ideals which approved an abundance of children for each couple, usually with the support of extended families; and rich and versatile environments which responded well to slow but continuous growth. In pre-industrial terms China's formal civilization was commensurate in scale and stability with her achievements at the agrarian, workshop and domestic levels, and poverty was not more marked than elsewhere. Much of the nation's effort under the People's Government has been turned to the search for means to restore China's standing to its old level, in the changed circumstances of the late twentieth century.

China as a Socialist Country

China is properly called the People's Republic of China (PRC). It has been ruled since 1949 by governments based on the Chinese communist party. Communist rule was the outcome of a long campaign, civil and military, which began in 1921 and continued throughout the Japanese war of 1936–45, and which was finally victorious in 1949. Taiwan island is the exception to this course of events. Taiwan was a Japanese colony from 1896 to 1945, and has subsequently been ruled as the 'Republic of China' by governments of the *guomindang* (*kuomintang*) party, which the communists overthrew on the mainland in 1949.

Chinese communism owes much to Soviet, especially Stalinist, precedent; but the Chinese communist party has not often been content to accept Soviet systems without alteration. There are two reasons for this. One is a strong sense of China's individuality, and even superiority, towards her neighbours; this seems to be a continuing legacy of China's historic isolation in east Asia and her long history of cultural superiority over neighbouring peoples. The other is the very marked differences between the human and physical characteristics of China and those of Stalin's Soviet Union – or even the Russia of the present time. The south-eastern half of China is its heartland, with 96 per cent of the country's people. These densely settled rural and urban populations occupy generally favoured subtropical or warm temperate environments, and the overwhelming majority are native Chinese speakers. The Russian heartland south of latitude 60°N and West of the Urals occupies only about one-eighth of 'Soviet' territory and perhaps one-fifth of the territory of Russia. Slavonic speakers comprised only about 78 per cent of the old 'Soviet' population, and by no means all of the Russian. 'Soviet' populations almost everywhere are much less dense, and natural environments much less favourable, than in China; and much of Russian (or Soviet) economic power and potential is based on mineral resources in the vast Siberian wilderness.

Finally there is potential for ethnic unrest in China, mainly in the central Asian provinces, but this cannot menace the existence of the state, as similar unrest in Europe and the Caucasus already does in the former Soviet Union. It is not surprising, in these very different conditions, that the Chinese communist party has often been out of step with Soviet ideas, or that 'the Chinese road to socialism' is often explored in the press. But Chinese socialist policy has by no means followed a consistent course since 1949, as will be shown;

and the 'Chinese road to socialism' seems to have only two essential characteristics – the assertion of China's intellectual and political autonomy, and the practical acceptance of each phase of communist party policy.

China as a Colonial Power

In spite of pressure from the Russian empire in the nineteenth century and later, modern China has retained a great part of the central Asian territory which it inherited from the historic Chinese empire. Almost half the republic's territory is of this kind, thinly populated by peoples many of whom do not speak Chinese, and whose civilization is different, often radically so, from that of the Chinese. 'Autonomous' status was supposed to defuse the possibility of ethnic hostility to Chinese rule in central Asia, but it has done so only very imperfectly; open ethnic tension has been slight except in Tibet, but the possibility of separatism based on ethnic differences remains. China's power in central Asia since liberation is based on two foundations – the immensely greater population of eastern China, and the capacity of Beijing to take effective administrative (and indeed military) action if trouble were to arise. It is not that most central Asians are 'oppressed', but they are not masters in their own territories. Beijing supplies aid to poor communities in Central Asia, and there are various kinds of subsidy. But it is argued in China that the sale of central Asian outputs further east at low prices much more than compensates for this.

China as a Less Developed Country

China of course ranks as an underdeveloped country or less developed country (LDC). But LDCs vary very much, and China fails to resemble the stereotype in several respects. Thus, unlike some other LDCs (for instance in Africa), China has a long history of civilization through most of which the level of development was the equal of, or indeed superior to, much of Europe – even in terms of technology. Chinese 'underdevelopment' resulted essentially from developments in Europe from the seventeenth century onwards, which created the dimensions and unprecedented dynamic of the modern world. For China 'the development of underdevelopment' has been a specific group of historic processes, which have owed much more to European and North American innovation than to Chinese backwardness (Lippit 1978).

Furthermore, most LDCs are not 'socialist' countries – that is to say, are not governed by a communist party. China has been so governed since 1949, but of course the development of China since that time has been unconventional – the Chinese communist party has had little contact with the Soviets since 1958, and communist policy in China has been independent and volatile. China between 1949 and 1979 did not usually accept aid for development – indeed in the 1960s and 1970s she gave aid which she could ill afford to clients in Africa and south Asia. China's uncompromisingly 'socialist' stance between 1959 and 1976 led some writers to take up the idea of 'learning from China' in the Third World; but the idea did not find many friends among Third World governments or ruling classes. Maoist guerilla movements did, however, develop in some Third World countries – the Philippines, Sri Lanka and Peru.

LDCs are usually marked by low income levels per person and per household, low living standards, lack of investment, poor infrastructure and often political systems with high levels of communal tensions. Some have also experienced incontinent population growth and excessive migration to the cities.

China answers to some of these stereotypes, but not all. Her level of income per person is one of the lowest on earth, much lower than Brazil's or Indonesia's or the figure for the South African black community. Along with India and Pakistan, she takes her place among the poorest 10 per cent of countries. Living standards in China are correspondingly low, though rarely disastrously so; and in some respects, such as health, standards are generally good. China has not lacked industrial investment since 1949; her infrastructure is in most ways adequate if not strong. Population policy has had its fluctuations, but during the past decade, China has proposed, and tried to implement, an exceptionally radical policy for birth control; and until recent years migration to the cities has been very efficiently controlled. China's low levels of income arise (like those of India and other Asian countries) less from low levels of production than from the immense size of her population, compared with the country's limited resource endowment. Policy since 1949 has exacerbated some weaknesses while eradicating others, but has succeeded in generating annual rates of economic growth which have usually been above 5 per cent.

Studying China

In the study of contemporary China, there are two related thresholds to cross. One of these is a kind of credibility gap – the problem of information, and how trustworthy the sources of information are in terms of factual accuracy, and also in terms of practical realism. The other is the problem of China's immense size. Western people often find China astonishingly uniformitarian, and even uniform, for so vast and varied a state, and find this observation hard to believe. Both of these are real problems, but neither need impede progress in the study of China.

A 'universal empire'

To take up the second problem first. Chinese provinces (figure 1.1) are of the same order of size and population as European nation-states, but Chinese official management, and domestic civilization among the Chinese-speaking population, are much more uniform than among the Europeans. Like other great Asian states such as India and Indonesia, China is neither a nation-state on the European model nor a sort of federation of such states: China has much more the character of a 'universal empire' based on a single many-sided scheme of civilization. This depends partly upon the origin of modern China as a slow organic growth through many centuries, including colonization movements adding territories little by little, partly upon management of the vast territory and immense community by a central authority and appointed bureaucracy, both under the People's Republic and through the traditional centuries under the historic empire.

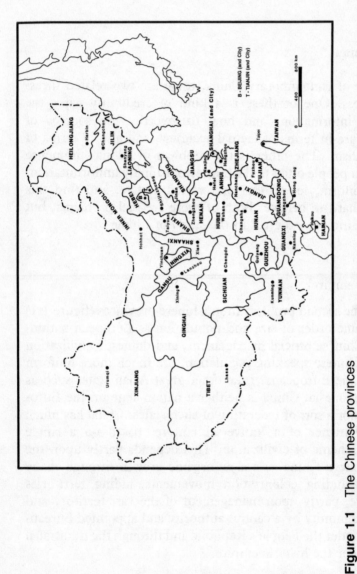

Figure 1.1 The Chinese provinces
This map is intended to introduce the Chinese provinces by name. These now number thirty, including the three great cities which have provincial rank, Beijing, Tianjin and Shanghai. Provinces in the densely populated east are much smaller than most of those in the central Asia hinterland; the historic empire had eighteen provinces, which did not include those of the north-east or of central Asia (except Gansu), or the three cities. Provincial capitals are also named. Taiwan is not part of the People's Republic, but Hainan is a recently designated province.

Information in China

The communist party system collects statistics with obsessive zeal, but like other communist parties, it is very secretive about them. Between 1957 and 1979 almost no comprehensive detail about Chinese geography, economy, resources, development or population was published; as late as 1977, when the total population was around 950 million, professionals in China were still using a figure of 800 million in published work. An abundance of local studies, especially from the countryside, reached the media during much of this long phase. They were distinguished by an untiring optimism whose main instrument was selective reporting. The media do not appear to have made a habit of untruth; the characteristic forms of misrepresentation were concentration on cases favourable to the current argument, and the simple omission of material which might be thought problematical.

The reasons for all this naive secretiveness are not very clear, but presumably lay somewhere between fear and contempt of criticism; perhaps realistically they were no more than simple force of Stalinist habit, ministering to Chinese official traditions of total authority. When reforms were introduced in 1979 and subsequently, it was recognized that if the people were to be drawn (as they must be) into the national reconstruction process, the state must take them deeper into its confidence. It is still rare to find published figures of more than local importance which have not already appeared in the national press, but the national press and the public are now provided with some very large quantities of data, especially the *Statistical Yearbooks* (*Tongji Nianjian*). But the characteristic forms of misrepresentation are the same as before, especially the omission of material which might be thought difficult to square with current policy. For a while in the early 1980s, when the Maoists could take the blame, there was an abundance of material about the inadequacies of the official system in the daily press, but in recent years this kind of material has given place to reports of official meetings and the like. Some of this concealment and casting of blame is not very different from what is attempted by Western political parties when in power; but of course in China the group in power in each phase can exert total control over publication, and where necessary does do so. Such control has no need to descend to matters of detail; editors know what is expected of them, and when circumstances change, they receive changed instructions. Some geographical topics, especially rural topics, are not generally treated as sensitive; others, especially urban topics and those relating to state industry, are much more so. When the state wishes to obfuscate, it often resorts to technical devices, such as the use of figures in non-comparable units (Chinese units such as *jin* or *dan*, instead of kilograms or tons), or expressed in non-comparable terms (values expressed in terms of prices of 1980 rather than the present, or indexes of a certain year's figure). Maps appear to be particularly sensitive; very few of the annual statistical volumes now published by provinces and even cities contain a map of any kind. There is still secretiveness, partly by habit, partly no doubt by intention, partly because junior units do not have authority to vary secrecy rules, and realistically must wait for senior units to propose innovations. But all this said, there is now much more and much better information to be had about China, both inside and outside the country, than at any time in the past. This situation, moreover, is not limited to factual material such as appears in the *Statistical Yearbooks*; it includes quite wide-ranging, intelligent discussions of policy in various fields, such as industrial and rural policy, the extent of poverty, the dangers of increasing pollution, weaknesses in Chinese industrial structure, and so forth. Use has been made of these discussions at many points in the pages which follow.

TWO

Resources and the Physical Environment

Abundance of Resources, Superabundance of People

China is a poor country, but the Chinese landmass itself is of subcontinental size, and its resource endowment must be considered, broadly, quite rich. The country's poverty arises from the imperatives of a limited body of land and resource occupied by a truly immense population. Much has been done since 1949 to widen the effective resource base and to limit the growth of the population, but of course not enough. Effort has also been put into the creation of new and more economical schemes of social management, but with only limited success. The Chinese perception of China as *di da wu bo*, a vast country with abundant resources, is not unrealistic in itself, but it must be qualified by recognition of the virtually limitless demands of both the population as direct consumers and the industrial system which it needs in modern conditions. Another older Chinese perception of China is more prudent – *di shao ren zhong* – the land is scarce and the people are many. In the present generation good land, forest, energy resources, water and other resources are all scarce; and the people are more numerous than ever before.

No single human system has tried to manage so many people before. Yet up to the present China is effectively still self-sufficient in food and most industrial raw materials, particularly energy. Twenty-one per cent of the world's people continue to be maintained in China from 7 per cent of the world's land area, and the same proportion of the world's arable land.

The Foundations of Environment

The detail of local topography in China is no less complex than that in other parts of the world, but the fundamentals of the physical landscape can be approached through some fairly simple generalizations.

A series of geological phases of mountain formation along various lines of weakness in east Asia have resulted in a mainly mountainous country with intervening basins, plains and plateaus (figure 2.1). It is these basins, plains and plateaus which generally form the nuclei of the main modern concentrations of river drainage, population and management – in fact, of the modern provinces. A typical east China province such as Hunan (basin) or Shanxi (plateau) represents one of these units. Typically, such a unit is around 150,000–220,000 square kilometres in extent, rather smaller than the United Kingdom or Alabama. Provincial units in western China are very much bigger. Figure 2.1 lays emphasis on the basins, which are the physical foundation of Chinese agriculture and, realistically, of the country itself; but of course it is the basins of the east and south-east which perform these functions, not the physically similar but climatically wholly different basins of central Asia.

As figure 2.1 shows, China has to be considered a mountainous country, with an official reckoning of 65 per cent of its land area comprising mountains, hills or plateaus. Moreoever, because Chinese territory extends deep into central Asia, in Tibet, Qinghai and Xinjiang, China includes some of the least hospitable physical environments on earth – sand seas and bare rock surfaces of great extent, and, in Tibet, some of the earth's most formidable mountain ranges, separated by rock platforms and deep gorges. Particularly in central Asia, the scale of these forbidding units of landscape is tremendous; but even in eastern China, important features such as the Qinling scarp south of Xian are on a continental scale.

The southern half of eastern China lies in sub-tropical latitudes; the northern half in temperate. The zone of transition is relatively narrow, located first along the line of the Huai river, then along the Qinling scarp which has been mentioned (figure 2.2). The Qinling scarp lies at about 34°N; the Huai a little further south. Broadly, to the north of the Qinling–Huai transition zone, January average temperatures fall below freezing, precipitation falls below natural evapo-transpiration, and soils become calcareous. These three qualities add up to a formidable difference in natural environments, especially for agriculture. To the south of the transition zone lies a relatively relaxed environment with adequate rainfall, a short mild winter and soils which are generally adaptable to rice cultivation. In this region, summers are generally long enough to permit full double cropping – for instance with a wheat crop which stands through the winter and is harvested in May, followed by a rice crop planted in June and harvested in October or November.

To the north of the transition zone, drought is always a threat, the winter is sharper and longer, and soils are both too loose to form paddy fields and chemically unsuited to rice cultivation. In the north China plain to the north of the Qinling–Huai zone, crops such as wheat can usually stand through the winter, but the summers are generally not long enough to permit full double cropping, and wheat can typically be grown (along with summer crops) through only one winter in two.

Western China is a wholly different world. It comprises about one-third of the vast and thinly populated heartland of the Asian continent, which China shares with the former Soviet Union and some lesser sovereignties such as Afghanistan and the Mongolian People's Republic. The northern half of this central Asian territory of China, centred on Xinjiang, Gansu and Inner Mongolia, comprises mainly basin and plateau country between 1000 and 2000 m elevation, with low rainfall (from the Atlantic) and very harsh winters. The indigenous population are Mongolian and Turkish speakers rather than Chinese, and traditionally nomadic. Many are Muslims, as in ex-Soviet Central Asia. The southern half of central Asian China lies mainly at more than 3000 m in elevation. It is centred on Tibet, but includes Qinghai and the western half of Sichuan as at present defined, and comprises a series of vast plateaus and basins oriented by mountain ranges whose highest peaks occur in the Himalayas, on the Indian frontier. The environments are generally extremely harsh, with low temperatures owing to the high altitudes, and limited agriculture in the valleys. Much of

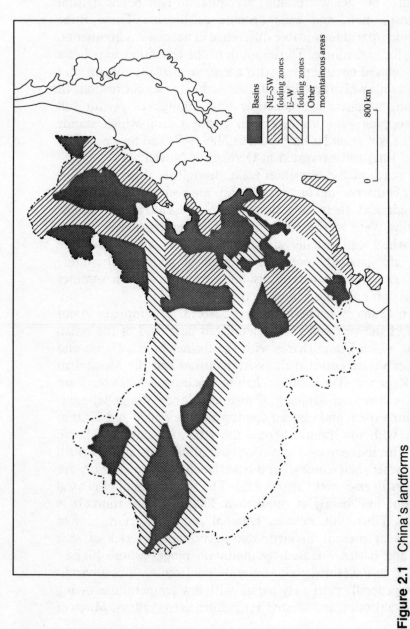

Figure 2.1 China's landforms
E–W folding zones intersect with others of NE–SW orientation, forming the characteristic underlying features of environment in eastern China. In the central Asian half of China, EW mountain ranges dominate. Broadly the plains are the main zones of human activity, but there are others as well, like the Pearl River delta in Guangdong.

Source: Atlas of Natural Geography 1984, 13–14

Legend:
- Basins
- NE–SW folding zones
- E–W folding zones
- Other mountainous areas

0 — 800 km

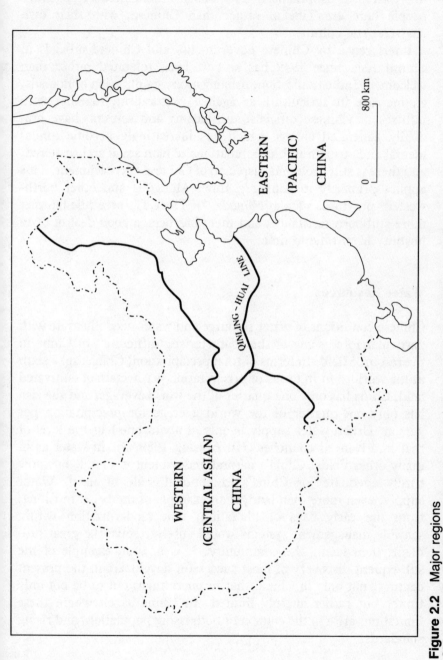

Figure 2.2 Major regions
China's territory divides into central Asian and Pacific realms partly according to elevation, partly according to climate. Pacific China divides at the Qinling–Huai zone according to winter climate, humidity and soils.

this vast area is practically or totally uninhabited. Indigenous people here are Tibetan rather than Chinese, with their own characteristic cultures.

Interference by Chinese governments and Chinese officials in central Asia since 1949 has at best been tolerated rather than welcomed. Industrialization, urbanization, stabilization of nomads, an increase in agriculture as against pastoralism, together with influxes of Chinese officials, technicians and settlers, have gradually generated change which has increasingly become fundamental and structural, but populations remain small and scattered, and there is still profound suspicion of Chinese encroachment. This applies primarily to Xinjiang, Inner Mongolia and other northwestern provinces such as Ningxia. In Tibet, Chinese rule has met more stubborn resistance, and there has been a good deal of open hostility from time to time.

Water Resources

Chinese conditions of water resource and water need illustrate with particular force some of the characteristic difficulties of China in the resource field. In terms of total precipitation, China ranks sixth in the world, but in terms of precipitation per hectare of cultivated land, China has only one-quarter of the world average; and she also has only one-quarter of the world average for precipitation per person. Urban water supply is only at about one-half the level of that in advanced countries (Hu Angang 1989, 3). In water as in many other fields, China's natural endowment is sound, but now barely adequate for China's exceptional scale of need. Water supply, even more than land, is an example of the belief in China, from the early days of liberation, that modernization within sensible management systems would surely resolve the great problems then facing the community. It is now an example of the subsequent discovery of most successful developers in the present century, not only in China, that resource turns out to be not only finite, but rather sharply limited. In China as elsewhere these limitations arise in the context of both rising populations and rising standards.

Figure 2.3 indicates the fundamentals of water availability. It is in effect a map for a moistness index, based on precipitation versus potential evapo-transpiration expressed in millimetres. Western China is very dry – much of it is desert territory. In eastern China there is a marked difference between north and south, with the zero line following the course of the Qinling–Huai line. Southern China has a broad surplus of precipitation, but the north is generally dry – not disastrously so, but to an extent which always conditions practical availability. In the north-east, happily, precipitation is greater and evaporation less, and drought is much less of a problem.

Irrigation is traditional in both north and south China, and of course in the oasis conditions of the west as well. The irrigated area in 1947 was 16 million hectares (*Ten Great Years* 1960, 130); by 1979 (the highest point reached) it was 45 million hectares, 56 per cent with powered pumps. In 1949 the great bulk of the irrigated area was certainly in the south, but in 1987 one-quarter of China's irrigation was contributed by the three provinces of the north China plain – Hebei, Henan and Shandong. In part irrigation water is diverted, especially but not only in the south; in part, particularly in the north, it comes from underground aquifers or new storage reservoirs. Some irrigation water no doubt helps finally to support underground aquifers.

Underground aquifers are the most problematic of the kinds of source mentioned above, and they are also critical in supplying many northern cities (Smil 1984). Beijing and Tianjin, with a combined population of 18 million people, are critically dependent on ground-water for all kinds of urban uses, both industrial and domestic, and for local irrigation. The same is true of a number of other cities in the north and north-east – Shenyang, Taiyuan, Xian and others. Cores of depletion have formed in aquifers at Beijing and several other cities as the water-tables sink, and chemical pollution is becoming serious in these reduced water-bodies. In the cities, both domestic and industrial uses have increased out of recognition in the years since 1949; urban systems of water use before liberation were virtually medieval in scale and character.

Not all Chinese water use, especially in the cities, is economical; one very odd middle-class enthusiasm (in a country full of cheap

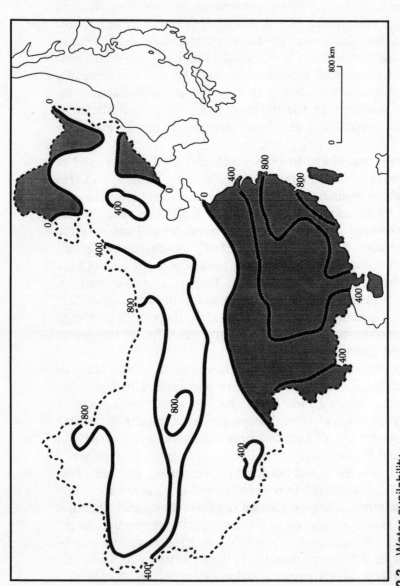

Figure 2.3 Water availability

The map shows water availability by setting precipitation against potential evapo-transpiration, expressed in millimetres. The shaded areas are those in surplus. The south has a handsome surplus of water, and there is also limited surplus in parts of the north-east, but over most of the northern and western three-fifths of China, precipitation is in greater or less deficit.

Source: Shang Chengyu and others 1986

labour but short of water and electricity) is that for washing machines. But indeed much modern industry uses water in very large quantities – 20–40 tons of water are used in making a ton of steel; 500–600 tons for a ton of fertilizer; 1200–1700 tons for a ton of artificial fibre (Li Jinchang 1988, 35). Meanwhile water pricing remains conservative, especially for state consumers, often not reflecting the rising costs of delivery of water, let alone the social costs of depletion, especially of underground water. Pollution is increasingly serious. The community discharges 34,200 million tons of used water annually into rivers and lakes, 82 per cent of it untreated. Some 19 per cent of surface water is now polluted, and in the dry winter, very seriously so (ibid. 49). Chinese systems of management are often casual and loose, which is tolerable even among dense populations when levels of consumer activity are old-fashioned and low, and its products few and biodegradable, but much less so in conditions of rapid economic growth. Modern chemicals in the water system self-evidently involve new dimensions.

Needless to say, Chinese water problems are regional problems. The dry north, especially the north China plain and its cities such as Beijing and Tianjin, is the main region of scarcity. Figure 2.3 indicates a clear divergence between a dry north China and a humid south. Water is not necessarily plentiful in the south, particularly in winter, but it is necessarily scarce in the north and particularly so in winter. Even the withdrawals which now supply Hong Kong in return for valuable hard currency, from the East River at Dongguan in Guangdong, are said to become difficult to satisfy in the dry winter.

There are means to handle scarcity of water, though they will not please everyone. More careful husbandry of water, including such measures as limiting the making of steel in Beijing, can be expected to yield significant dividends. However, the Chinese communist party is prone to a preference for big-scale reconstruction schemes, rather than improved housekeeping. In this field such a scheme (or group of schemes) is already under discussion – the diversion of water on a grand scale from south to north (Yu Chengsheng 1985). This project is for diversion of water from the Danjiangkou reservoir on the Hanjiang river in Hubei, northwards across the low watershed south of Zhengzhou, across the Yellow River and finally to Beijing. It is still under discussion, but seems likely to be undertaken, later if not sooner. But it is unrealistic to imagine that a big scheme of this kind can solve all the problems of access to water, and still less those of waste water and pollution. Continued population growth, continued rises in standards of living and hygiene for the mass of the people, and continued growth of manufacturing, must all put mounting pressure upon both water delivery and discharge systems, and upon the water resource system as a whole.

Energy Resources

Energy resources are of particular interest throughout the world, and those of China are no exception. Here too China's problem is less lack of resources than excess of demand. Although China is rich in energy potential by most national standards, she is much less rich in energy per person. The 770,000 million tons of accessible coal reserves amount to only 700 tons per person of the population. The projected energy output by year 2000 of around 1300 million tons coal equivalent will amount to only about one ton per person, 40 per cent lower than the present world average (Li Jinchang 1988, 17).

The geographical distribution of China's energy resources is also somewhat problematical. Most coal and almost all oil are in northern or north-east China; the eight southern provinces have very little of either. Natural gas is mostly in Sichuan; water-power potential mainly in the south-west, and (understandably, in China as in other major countries) mainly in remote mountain regions. In addition, high-grade energy resources are relatively scarce and the costs of exploitation can be high. Finally, it has to be remembered that commercial energy sources like coal are still very scarce in most parts of the countryside. Rural China depends mainly on plants for energy – it consumes 180 million tons of firewood and 230 million tons of straw annually, to the serious loss of the rural environment on both counts (ibid. 18).

There are also operational weaknesses in the energy industries – typically, infrastructures which are overworked and underserviced, and production systems based on short-term considerations. It is not necessary to sympathize with all the state's problems in this respect. Investment in the manufacturing side of heavy industry has been abundant (some would say excessive) for decades; but the primary and tertiary needs of industry have been badly neglected.

Coal

China is very rich in coal. Her resources (which are still not exhaustively surveyed) rank third in the world after those of the former Soviet Union and the United States, and in recent years her coal output has been first in the world, around 20 per cent of the world total. Coal is widespread in China – some 60 per cent of the 1900 or so counties have coal deposits of some sort. Chinese coal is also usually of good quality, with a wide variety of types, and conditions for extraction are usually good, with most depths less than 500 m. The coal economy is, however, subject to various limitations, some of which are expressed in production patterns. Around 60 per cent of proved resource is located in Shanxi or Inner Mongolia, and this includes much of the most accessible resource. Shanxi's position as the country's main coal producer has actually strengthened since 1982 (from 22 per cent to 26 per cent in 1989), and according to present plans seems likely to go on doing so. At the other end of the spectrum from Shanxi, the greatest proportionate increase in production since 1982 has been the 82 per cent increase in Guizhou, which together with Sichuan and Yunnan may have the second largest group of reserves, and by far the largest group in the south. Other major increases in this phase of rapid growth (total output increased by 58 per cent in 1982–9) have been in Heilongjiang (65 per cent), Sichuan (73 per cent) and Yunnan (69 per cent) – the last from a very small base (CSY 1990, 435, 445; SYC 1983, 262). Failure of output to increase proportionately has been most marked in the provinces of the north China plain and its peripheries, which also have a major group of reserves, and where (in Hebei) modern coal-mining began its rise. In the extended group Liaoning–Hebei–Henan–Shandong–Anhui, all important producers where in some cases (Henan) there have been important new coal developments, output increased by only 34 per cent in 1982–9. This seems to reflect the disproportionate need (both geological and environmental) for costly investment in new developments in the north China plain, rather than in Shanxi where much recent development has been open-cast. But in terms of location and the practicalities of transportation of coal, Hebei, Liaoning and other northern provinces are necessarily superior to Shanxi, because they are much closer to the concentrations of demand represented by industry and population in the north China plain and its cities, and also to Shanghai.

In 1952, the coal outputs of Hebei, Liaoning and Shanxi were very close in size; each produced about 11 million tons in a total of some 66 million tons (*1949–84 Statistical Materials*, 53). At that time Henan–Anhui produced only some 6 million tons, and the whole south-west only 4 million tons. Heilongjiang, however, already produced around 6 million tons in 1952. Of all provinces, the proportionate increase in production since liberation has been greatest in Fujian, with an increase from 2000 tons in 1952 to 5 million tons in 1983 (ibid.). This figure had risen to 9 million tons by 1989 – still a small output (1 per cent), but strategically located in the south-east where coal is scarce. Lacking provincial data from the 1960s and 1970s, we have no clear indication of the phase in which abundance of resource (in Shanxi) and regional need (in the south-west) came to prominence in the geographical pattern of the Chinese coal industry; presumably they did so progressively after 1962. Important phases of Chinese planning principles, as well as the need for minerals, suggest the progressive relocation of Chinese industry away from the coast, but industrial inertia remains strong, supported by vital practical considerations such as the location of skills in the workforce, the need for employment and higher productivity in older industrial areas.

Problems in the coal economy

Year-to-year management of the immense coal business is complex and in some ways problematical. Coal is usually scarce within the community, sometimes critically so; the coal economy is not prosperous, and the coalfield communities are not satisfied with the system. A series of serious problems have been identified (Zhu Shuofu 1989; Liu Xieyang 1989).

Demand, including planned demand, is rising too fast. Demand for coal foreseen in the five year plan for 1990 was some 10 per cent higher than supply foreseen in the same plan. This rapidly rising demand is mainly in the state's heavy industrial sector, or in electricity generation. Transport of coal, which is mainly by railway, is always difficult; at times stocks even in places like Shanghai are down to a few days' supply.

Official investment in the state's coal industry has been inadequate, in some phases falling below what had been promised. Low investment over a long period of years has had cumulative effects. Official coal prices are far too low; as a result, state subsidies are required to offset losses. This no doubt tends to discourage state investment, which cannot be profitable when prices are so low; it is also constantly discouraging to the local coal administrations. It is also one reason why miners' pay and motivation are low. For the same reason, labour conditions are poor, and tending to fall further behind even the present low standards. Inevitably, when coal prices are kept low, a 'scissors' gap opens up between the value of coal production and the costs which production must meet. Between 1980 and 1989, steel prices doubled relative to coal, and timber increased by six times. Levels of mechanization in the mines are low – around 30 per cent; and output of coal per worker per shift remains remarkably low at 1.2 tons (*CSY* 1990, 455). Little progress has been made in either of these respects since 1979 or indeed since 1952. Coal-washing facilities remain limited; capacity is only about 15 per cent of total output. Most kinds of energy use in China, especially coal, are wasteful by international standards.

Perhaps surprisingly, 90 per cent of recent increases in coal output have been contributed by township and village mines, operated by local groups with only dubious legal status (because in China minerals belong to the state), and sometimes at the cost of inconvenience, loss and even damage to state mines in the same areas (Sun Hao 1986). From 20,000 in 1978, small mines rose in number by 1985 to 60,000, mainly in the years 1983–5; by 1984 their output was 50 per cent of China's total.

In that phase, due to the increase in output from local mines, prices of non-state coal fell for a while, but later revived. Understandably, informed opinion in China is divided about the desirability of expansion of the township mining industry. Its wastefulness, competitiveness with state mining, and indifference to safety procedures suggest contraction; its capacity to respond rapidly to demand suggests further expansion. Among other advantages, its economical operation appears to make it responsive even to low coal prices, and its independent status enables operators to negotiate for advantageous prices in conditions of general scarcity of coal. According to the Minister for Coal in 1983, small mines are expected to produce up to 500 million tons of the proposed output of 1.2 billion tons by the year 2000 and existing state mines up to 400 million tons; up to 400 million tons should come from new mines, of which about one-half will be open-cast (Lu Longwen and Chi Tingxi 1983).

Most small, locally run mines are in Shanxi province, peripheral to the state's main coal resources. Shanxi has not been happy about its status as a prime 'coal base'. Arguments are used not unlike those adopted for agriculture by provinces such as Hunan, which are expected to supply grain cheaply to urbanized coastal regions which are already richer and enjoy much better economic opportunities. Too little of Shanxi's coal, it is argued, is retained in the province to provide the foundations for an industrial economy. It is not yet clear that the mushroom growth of township mines in the 1980s has affected this argument; in fact local enterprise seems to be more interested in mining than industry. Meanwhile Shanxi authorities point out that the most accessible resources are being exhausted, that mining costs are rising fast, and that workers are beginning to leave mining for jobs in the 'free' commodity economy. Meanwhile, local use of coal in Shanxi is said to be very extravagant.

As time goes by, it is increasingly clear that China's main coal deposits are quite localized. Figure 2.4 shows the area of the 'energy base' designated in 1985. This area covers only 12 per cent of Chinese territory, but more than 70 per cent of Chinese coal resources. It is located fairly centrally as far as northern China is concerned, but of course at some distance from the coastal cities which are the principal consumers – from Shanghai to western Henan is around 800 km; further still to Shanxi.

New developments are still taking place in coal exploitation, such as the new Shenfu coalfield in northern Shaanxi, in poor loess country between Shenmu and Fugu, which is intended to supply Chinese coastal cities with steam coal and to provide coal for export.

The main coal port, for both export and shipments to south China, is Qinhuangdao, east of Beijing. Qinhuangdao is now linked directly to the coal city of Datong in northern Shanxi by a railroad intended primarily for coal. There is talk of a new coal port and connecting railway from Shenfu.

Oil

The status of China's oil resources is very different from that of her coal. China produces about 17 per cent of the world's coal, but only about 3.4 per cent of world oil – ahead of British production, but not very different. Almost the whole of the vast output of Chinese coal is consumed in China, but of oil and petroleum, some 20 per cent of the total output of 138 million tons (1989) is exported, mainly to Japan. High international oil prices have naturally tempted Chinese officials to sell oil abroad whenever possible; in 1987 (*CSY* 1990, 603–4) oil and petroleum ranked alongside garments as one of the two most valuable groups of Chinese exports, and contributed 7 per cent by value.

Chinese oil consumed at home, however, is not as different from coal in use as might be expected. Less than 20 per cent is used in transport, no doubt owing in part to official control of supplies, in part to lack of vehicles – although this figure may be raised to around 30 per cent if allowance is made for the widespread use of farm tractors in rural transport. Significant quantities of oil are used in the oilfields and refineries (13 per cent) and the chemical industries (16 per cent), but more than one-half of Chinese oil is used in ways akin to those for coal, in the power generation, building materials, smelting and machinery industries (*CSY* 1990, 468). It must be the aim of policy to reduce this kind of use as far as possible, and to substitute the use of oil to produce motion or chemical raw materials, for its use to produce heat.

Oil in China comes from a number of fields, of which Daqing in Heilongjiang, discovered in 1959, is still the most important. A long-distance pipeline links Daqing with the north China plain. In 1989, Daqing produced 40 per cent of China's oil (*CSY* 1990, 445). Most of the rest was produced by onshore installations on the coast of the Bohai Gulf around the mouth of the Yellow River, at Shengli

Figure 2.4 The 'great coal base'
The shaded area is approximately that designated officially as China's coal base. It has 12 per cent of China's area but 70 per cent of its coal reserves and 42 per cent of coal output; and it supplies 90 per cent of China's commodity coal

800 km

0

and Dagang, a field discovered in 1962 but exploited mainly in the 1970s. From this field, Shandong and Hebei together produce about 28 per cent of China's oil output.

Costs are high and productivity is low in the Chinese oil industry, but oil prices in China are very low – about one-quarter of world prices. The price of oil is now not sufficient to cover costs. As a result, the industry is in debt and depressed, which much less capacity to invest (it is argued) than is necessary to keep exploitation up to date, and to find and develop new wells. At the same time, no exceptional new discoveries are being made, in spite of extremely flattering estimates of 'potential' reserves; people in the industry point to the limitations upon investment (Li Yongzeng 1989).

An Energy 'Crisis'

In the field of energy, there is now talk in China of a 'crisis' (Li Zhisheng 1989). In the short term, this idea seems to be based on two kinds of scarcity: of coal and of electricity. Each is the result of both sluggish supply and rapidly growing demand – and of course shortage of coal at the power stations is one reason for lack of electricity. Electricity is expensive for ordinary consumers, but it is less so for state factories, and it is often used extravagantly by many kinds of business, especially hotels and some factories, with air-conditioning and lavish lighting. It is not uncommon for power to be shut off for days together, even in Shanghai, or for regular cuts to be made on particular days.

The crisis of demand is also, of course, a crisis of supply. Heilongjiang, the northernmost province, is China's principal supplier of oil, and third (after Shanxi and Henan) of coal. In the decade from 1980 to 1989, Heilongjiang's formerly prosperous industrial economy has fallen heavily into debt, due to prices for outputs which are low, and which have risen much less than prices (mainly through the state) for imports. Around 70 per cent of these losses are in the coal and oil economies owned and managed by the state (Li Jie and Liu Xin 1990).

Clearly the problems do not begin and end with those of consumers. There are also long-term issues involved – failure of coal outputs to grow at the same speed at state industry, let alone consumer and non-state industrial demand; failure of electrical installations to meet demand; failure of Chinese energy users to adopt economical practices; failure of the state to take the needs of the energy industries seriously. Meanwhile population, production and business continue to grow, and consumer standards to rise. The potential for further rises in all these respects is virtually without limit. Energy is one of the several respects in which the past decade of attempted conversion of China into a modern country has revealed a sharp divergence between social expectations and the capacity of the present resource exploitation systems to satisfy them. Chinese use of energy is nevertheless low by world standards, even while coal output has been pushed up to the world's greatest.

Minerals Apart from Fuels

Apart from fuels, mineral resources belong to three main groups – those involved in making artificial fertilizers, metal ores and building materials. China is a very large country, and in the nature of things contains a wide range of resources of all three kinds – but with the usual provisos of a distribution which does not parallel that of demand or fuel for smelting, and deposits of different minerals which differ greatly in size. Thus, materials for potassium fertilizers occur mainly in the far west. A few minerals are scarce in China, notably ores of copper and nickel, and a few are particularly prominent, notably ores of tin, tungsten and antimony, of all of which China is a prominent exporter. Outputs of aluminium are likely to increase when there is more hydro-electricity. Iron ore is mined in many parts of China on a small scale, but production is on a big scale at various places, mostly inland, where iron-and-steel plants have been established, such as Maanshan in Anhui and Baotou in Inner Mongolia. China does not usually either import or export iron ore.

Accessibility in China

The transport network inherited by the People's Republic in 1949 was essentially a strategic one, based upon a very limited rail system, much of it single-track, linking most of the provincial capitals. Distances are great, and on many lines engineering difficulties are daunting in the extreme; everywhere inland from about 113°E watersheds are over 1000 m in elevation, and often precipitous. Modern roads were few and narrow. The rivers, as throughout traditional times, carried a great deal of the country's traffic; but they too had problems, such as portages between headwaters, the long haul of ships upstream by gangs of coolies through the Yangzi gorges, the scarcity of water in many rivers in winter (particularly in the north), the flooding of summer, and the then derelict condition of the only major north–south waterway inland, the old Grand Canal between Hangzhou and Tianjin. Archaic transportation systems were one feature of Chinese environments which lent an air of necessity to Maoist insistence on self-sufficiency everywhere.

Construction of railways and roads has not been a first priority since 1949. The strategic rail network has been strengthened by a number of important lines, some involving formidable engineering achievements, as between Baoji and Chengdu (from Sichuan north through the Qinling to the Yellow River basin – in the 1950s), Guiyang to Kunming (in the extreme south-west – around 1960), and Beijing to Taiyuan (in the 1970s). In all, and including branch-lines and double-tracking, some 30,000 km of railway was built between 1950 and 1980 (Sun Jingzhi 1988, 326–7). This was not an achievement suggesting any special distinction and in fact even in eastern China (apart from Liaoning) the network remained primarily a strategic one. It remains so still, since there has been very little priority for transportation since the reforms – only about 4000 km (7 per cent) of growth in the period 1979–89 (Leung 1980).

Roads have fared a little better since 1979, with 138,000 km (16 per cent) growth in the period 1979–89. But China has only begun to encounter the general world (including Third World) experience of motor traffic during the past decade, and up to the present there is little understanding of its potential either for growth and economic and geographical transformation, or for social nuisance and disaster. Traffic is still limited on China's roads, but transport businesses can be very profitable as well as difficult for the officials to supervise; drivers are often over-confident, roads are old-fashioned and accidents are frequent. About one-half of Chinese passenger-kilometres are carried on the railways, and most of the rest on the roads, with air and water traffic sharing about 7 per cent. About 40 per cent of freight traffic uses the railways, but 44 per cent the waterways (*CSY* 1990, 499, 502–3).

Accessibility, however, is a little more complicated than simple statements about the transport systems. There are several levels in China, as in other countries. For the officials, even local officials, accessibility is not normally a problem – for long distances they fly, for shorter distances or those for which flight is not practicable, they travel by chauffeur-driven car – officials appear never to drive themselves, and the official priority demand for this scarce skill cannot but hinder the growth of motor traffic available to the public. For ordinary working people, accessibility has much more to do with local transport systems between village and town, and between local town and county capital or major city. There is a good deal of evidence that at this realistic level, road transport (often only by bicycle) is the heart of the matter, and that as far as waterways are concerned, bridges or ferries are quite as important locally as the waterways themselves are for long-distance transport. Individual localities each have their own experience with bus services, ferries and road construction.

For the minority of independent businessmen (*getihu*) who operate buses and other kinds of transport systems (including motor-bike pillions which collect passengers from bus stations, as in Thailand and Indonesia), local transport can be highly profitable, mainly because it is still scarce. Local transport indeed operates within systems of scarcity – of vehicles and fuel, and of course of official support and protection (stories have appeared about official bus enterprises persecuting *getihu* bus enterprises; Jin Yong 1987). It is no doubt by reason of these scarcities and the general lack of transport in the countryside that good money can easily be made in this field by those who exploit it. There is a long way to go in China before a rural town of 40,000 people in a densely populated countryside can have a bus or minibus arriving or departing every minute during business hours, such as we can find in Java or (according to report) in southern Nigeria.

Rural marketing areas are a feature of special importance. The general wisdom in China is that peasant marketing takes place in cities and big towns from within a radius of about 20 km, but for small towns, within a radius which realistically is about half the distance between such towns, often as little as 2 or 3 km in densely populated areas. For such purposes, peasants use tractors, bicycles or buses – but the big public buses often object to passengers carrying produce to market; *getihu* minibuses are more accommodating. An area of the latter kind would be around 60 sq. km in extent, with a population of around 5000 rural households.

The Wild Environments and the Forests

In China the natural environments are usually considered to be intensively humanized, and wild environments consequently scarce. But in fact only about 10 per cent of China's surface is cultivated. Forest and woodland (not necessarily first-growth) occupies 13 per cent, and open grassland (not necessarily 'natural') one-third of the total. More than (58 per cent) of the country lies at more than 1000 m elevation, and one-quarter at more than 3000 m (*CSY* 1990, 5).

Central Asia

Needless to say, the first distinction to be made in a review of the wild environments in China is that between the central Asia mountain ranges, plateaus and basins, and the eastern half of the country (figure 2.2). Most of the central Asia environments are quite untouched by man, partly because man is a scarce phenomenon over most of this area, and partly because these environments (sand, seas, rock surfaces, mountain ranges, with very low rainfall and intense cold in winter) offer nothing at all to human need. This of course is the foundation of the scantiness of the populations.

In Xinjiang, mountain chains like the Tian Shan receive more rain, and streams flowing out from the mountains nourish chains of oasis settlements. These chains of settlements also represent the caravan routes of traditional times, by which China maintained its limited but sometimes momentous contacts with the rest of Asia. They are also the heart of modern development in Xinjiang. In Xinjiang too, forests occupy many mountain environments, with vegetation communities which differ according to elevation. However, the survival of forests in modern conditions in Xinjiang, as elsewhere in China, depends greatly on the protection given by isolation. The historic means of livelihood in most of central Asian China was nomadic herding, and most of the open grassland lies in this region. Some environmental decay has certainly taken place in the grasslands during historic time – it is not clear how widespread is this decay, or to what extent localities differ. Forest was certainly more widespread in earlier centuries. One central problem in these harsh, dry environments is the lack of regenerative capacity in the vegetation; but of course the great engine of degradation is man, with his animals and his appetite for timber.

Understandably in view of the harsh physical conditions, the wild biological resources of central Asian China are not rich, in terms of either plants or animals, even without the destruction created by man. In short, central Asian China is a group of the most difficult environments on earth.

Pacific China

In terms of environments, central Asian and Pacific China have little in common. The latter is in most respects a highly favoured region. Up to the late eighteenth century (with total population around 300 million in 1800), the environments must have appeared virtually limitless in scale, diversity and even generosity. The natural vegetation in all areas (except possibly the loess) is forest; but the forests of north and south differ very much, due to differences in degrees of drought and cold in winter. Forests in both areas are very varied, however, and throughout very rich in species, including species not found elsewhere. Inevitably, these rich forest habitats have been subjected to human pressures for many centuries, and, during the twentieth, to pressures from which recovery is increasingly difficult.

Forests of the kind described offer many ecological niches favourable to animals of all kinds, both vertebrates and invertebrates; and China's natural systems are also very rich in animal species. But here too human interference, probably conjoined with climatic deterioration (in the sense of dryer and colder) since Song times around AD 1000, has tended to gradual squeezing of species into narrow enclaves – for instance of elephants into a few valleys in Yunnan, from a territory which in ancient times extended as far as the Yangzi.

Timber and bamboo are the principal outputs of the remaining forests. Bamboos of all kinds are of particular interest, with many uses (furniture, tools, building, baskets – in traditional times even ropes were made of woven split bamboo). The varied flora and fauna have very varied human uses – food, traditional medicines, gathering honey from wild bees, and so forth. Trade in many such items between north and south in China has a long history.

Progressive loss of the forests stands as a grim threat to Chinese civilization itself; but most rural fuel is still firewood, taken from forests and woods where these still exist, and afforestation is still limited both in scale and in its capacity to recreate the varied complexes of the natural forests.

Pollution

In Mao's time, pollution was not prominent in China. It is true that state factories burning coal emitted clouds of black smoke, that state mines sometimes polluted water supplies downstream, that urban cleansing services remained limited, and that fertilizer and pesticide might be applied over-enthusiastically. Many of the examples given by Vaclav Smil (1984) relate to the 1970s. But in those phases there was little consumer production and (for that reason and by reason of poverty) little refuse; drinks cans were few and for that reason kept for re-use; most fertilizer was natural waste from people or animals; industrial production outside the cities was scanty; and traffic in town and countryside was light. During the reform phase since 1978, pollution has increased to a striking degree in very many forms and, it would seem, for several different reasons. One reason is the increase in personal consumption of all kinds of items, from timber to plastic bags. This relates to heaps of refuse which are now increasingly appearing in both town and countryside. Another is the increased use of chemicals by farmers, which contributes markedly to rural pollution of soil and water. At the same time, rural lavatories are now emptied less frequently. Township industry now adds its growing capacity to turn out mountains of rubbish. Traffic is now much heavier both in town and on the main roads; accidents are frequent and can be devastating. In town, people no longer stay at home after work as they did in Mao's time; they walk about and throw down peanut shells and melon skins – like rural people, they often have little sense of a need for tidy habits. Increases in velocities and densities of circulation have led to crowding everywhere, and the same is true of continuing population growth (305 million, or 38 per cent, between 1969 and 1989). Mention has already been made of the pollution of water resources. The central problems turn out to be those which are already familiar as problems in resource use – rising populations and rising standards of living among China's vast population.

Turning to broad considerations of the environment, conditions are bleak in many fields (Xiao Jiabao 1989). Soil erosion is still increasing. About 5 million tons of soil is lost annually, including plant nutrients, an amount similar to China's annual output of chemical fertilizers. Forest areas are still slowly falling, as cutting continues to exceed growth. Sand seas are still encroaching on farmland along the northern edge of cultivation in provinces such as Shaanxi. Eighty per cent of Chinese sewage is still discharged untreated into rivers. Soil pollution by pesticides is a growing menace. Clearance of mountain forest and draining of lakes for the sake of new arable land has certainly led to increased risk of floods such as those of 1991 in the Yangzi provinces.

THREE

The Historic Foundations
of Contemporary Change

Origins and Foundations

The People's Republic inherited its territories and peoples from the Republic of China of 1911–49. This 'bourgeois' republic overthrew the historic Chinese Empire in 1911, and in its short and violent history experienced the long resistance war against the Japanese (1936–45) which merged with the Second World War, and the even longer accompanying civil war against the communists (1926–49, in various forms), as a result of which the communists were finally able to seize the government. The historic empire, founded or refounded in 221 BC, was by far the world's oldest state structure with a continuous history.

The Chinese Empire, however, and Chinese civilization itself, was by no means as old-established as the ancient states and civilizations of Egypt and the Middle East; its hallmark has been continuity from its own beginnings, rather than extreme antiquity. But continuity in both domestic and official civilization, extending through the past thirty centuries at least, has been the source of much conflict and suffering, as well as great underlying stability and confidence.

Recognizable Chinese civilization is usually traced back to the second millennium BC, and in terms of regional origin to the middle section of the north China plain, along the Yellow River, in modern north Henan, south Hebei, west Shandong and south Shaanxi. This civilization belongs to the bronze age. There is evidence from the archaeological record that these ancient people were of similar physical type to modern north Chinese, and that their language was similar in structure to modern Chinese, and written in 'characters' akin to those of modern times. Bronze age China already had substantial cities and a powerful official bureaucracy, but it was primarily an agrarian community growing millet and beans for food. It occupied an environment with much more forest and marsh, and almost certainly a warmer and wetter climate, than modern north China.

By the time of Christ, under the Han dynasty from which the common term 'Han', meaning Chinese, derives, most of the characteristics of the civilization of the historic empire were already established. Han was an iron age state with a population of around 60 million. It was still overwhelmingly agrarian, and already had a powerful central government and wide-ranging official controls. State contacts had already widened to include most parts of the south, but the civilization of the Han people was still confined to the north China plain and its peripheries. Farmland was already crowded. The conventional landholding was 100 *mu*, around 4.6 hectares, but many holdings were much smaller, and under Han there were widespread complaints of the rich monopolizing land-holding – the common people, it was said, 'had not land enough in which to stick an awl' (*Han Shu*). Millet porridge was the staple food of the countryside, as it continued to be in northern China until the twentieth century, but by Han times wheat was also grown, sometimes as a winter crop to be harvested in late spring. The rich ate meat, often game – not all the wild land was yet put under cultivation; but poor people kept pigs, dogs and chickens for food, though for sale rather than to eat at home. Needless to say the life of a farming family was hard. According to an ancient writer, 'In springtime they cannot avoid wind and dust or in summer the heat; in the autumn they cannot escape the rain, or in winter the cold. They have no days of rest' – but this writer goes on to point out the privileges of the commercial classes, in spite of their low status under the law. 'They do not have the hard work of farmers, but they get hundreds and thousands of cash' (*Han Shu, Shi Huo*). Communist party hostility to private business in Mao Zedong's time and even in recent years can call upon strong traditional allies in China.

Neighbours of China to North and South

The centuries which followed Han produced long phases of secure government and prosperity, as in the seventh century under Tang or the eleventh under Song. They also produced important phases of weak government, invasions from the north and widespread destitution. Traditional China had neighbours of very different kinds to north and south. To the north, on the central Asia fringe, her neighbours were peoples such as the Mongols and central Asian Turks, many of them nomadic pastoralists with uncompromising military traditions. These peoples domesticated the native horses of the steppes, and horses gave them great mobility in war. From Han times onwards, the settled farming community and management systems of the Chinese were always at risk of attack from some union of nomad tribes; the Great Wall, which purported to define the northern frontier, was rarely a practical deterrent to a serious attack. To the south, on the other hand, China's neighbours were farming people, settled generally in small local groups. These peoples were akin to those of modern south-east Asia rather than to the northern nomads – 'Tai' groups among them were related to the modern Thais of Thailand. Among these politically and militarily weak peoples, the Chinese state was powerful and willing to exercise the power it had to expand its territory. Continuous Chinese expansionism in the south is the obverse of frequent Chinese impotence in the face of nomad aggression in the north.

In traditional times, both aggression and impotence had farreaching consequences. In a number of important phases, nomad leaders found it possible to conquer most or all of northern China, and to establish themselves as Chinese emperors for generations. Under Kubilai Khan, exceptionally, the Mongols succeeded in conquering all China and ruled for 60 years as the Yuan dynasty; and the last of the northern dynasties, the Manchu Qing whom the Republic overthrew in 1911, had ruled all China since 1644. In the process of ruling China, nomad dynasties and gentry families inevitably became sinicized, though that did not necessarily reconcile their Chinese subjects to them.

Warfare in the north was one of the factors which stimulated Chinese interest in the south. In the centuries immediately following Han times, when the empire was partitioned and internal warfare spread, the important tradition of southward migration of ordinary farming families was established. At first the 'south' of the migrants was the basin of the Yangzi, central China by modern reckoning; but in subsequent centuries families moved further south. At first, no doubt, the vast abundance of land, very thinly occupied or totally empty of settlement, encouraged migrant occupation – though animal and insect pests, especially mosquitos carrying malaria, were formidable in what was basically a jungle environment. Settlement by families was accompanied, and sometimes preceded, by occupation by Chinese administration – officials, army detachments, the possibility of taxation. Already before Han times Guangzhou (Canton) was a forward part of the empire; little by little the intervening subcontinent was claimed and occupied. For many centuries under and after Han, North Vietnam was administered as part of China.

It is not too much to say that the historic empire was created by the land colonization movements and the accompanying extension of official management and government. Figure 3.1 shows stages in the whole process, though it cannot allow for degrees of intensification of Chinese occupation. By AD 1000, the broad shape of the historic empire was established, including the long salient into central Asia, mainly in Gansu province, which already existed in Han times. Later settlement was mostly in much more difficult country in the south-west and on the northern frontier, and of course much of it was very thin. Taiwan and Hainan were almost afterthoughts, but Taiwan particularly has become a major contributor to the rich variety of Chinese civilization.

Movement south, in spite of the insects, was generally movement into a better natural environment. Winters in the south are shorter and warmer, and rainfall is much more reliable; soils are sticky and water-retentive and so suitable for rice, the preferred grain in China as in most of east and south Asia. Not only crowding and hardship at home in the north, but more and better land ahead in the south, encouraged the millennial migration of Chinese families. By Song times (around 1100) the great riverine plains of the south were quite densely peopled, and the earlier heavy population densities of the north China plain had somewhat relaxed; by 1400 under Ming, the Yangzi valley and the south had taken up the position of primacy in population which they have occupied ever since. By that time, too, the centre of gravity of China had moved out of the north China plain and to the area at the mouth of the Yangzi, in southern Jiangsu and northern Zhejiang and centred in modern times on Shanghai, which the Chinese call Jiangnan – 'south of the river'.

The assimilation of the south during the first millenium AD much more than doubled the resource base of the empire, and endowed the Chinese community with one of the richest farming environments on earth. Movement south was movement into progressively warmer, wetter and more varied environments. Because wetter and more varied, these environments were also less fragile than those of the north. Forest and bamboo remained on the less accessible mountain slopes, cleared only gradually for timber or firewood – some remains up to the present. Traditional land use, with its limited use of pastoralism, made little use of cleared hillsides, but the flooded paddy fields in the valleys were an important check on erosion lower down in the water system, if not on the mountains themselves. Traditional crafts from the north, in textiles or boat-building or metallurgy, were revitalized in the warm and generous environments of the south.

There are important reminders of the long colonization process. At present, in about half of the southern provinces, minority peoples exist who represent the pre-Chinese inhabitants. In the 1950s, following Soviet precedent, most of these minority populations were recognized and acknowledged by the creation of 'autonomous' units of various kinds, from 'regions' at provincial level (Guangxi) to autonomous counties in provinces like Yunnan and Sichuan. The same system recognizes the generally much larger non-Chinese groups of the north, as in Inner Mongolia and Xingjiang Uighur autonomous regions. There are also Kazakh and Korean units of the same kind, and in Ningxia the 'Hui' (Muslim) autonomous region which recognizes not only religious tradition, but an identifiable and important social group. Minority groups now enjoy some exclusions from the population control policy and have often experienced less formal control than the Han populations, sometimes perhaps because of isolation in mountain areas; but they have never enjoyed realistic 'autonomy' under the People's Government. Figure 3.2 shows the distribution – occupying more than half of the Republic's area – of people now speaking minority

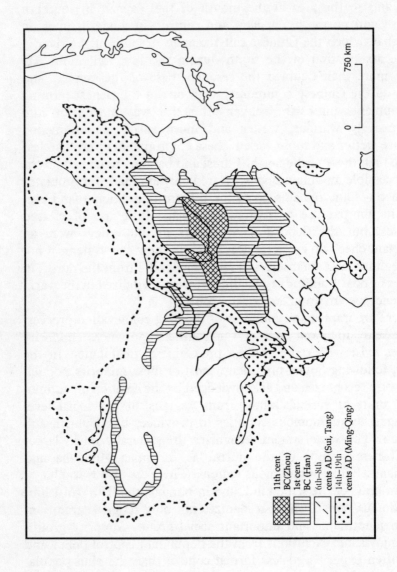

Figure 3.1 Phases in the Chinese occupation of China
Source: Adapted from Geographical Research Institute 1980, 55

Legend:

11th cent BC(Zhou)

1st cent BC (Han)

6th–8th cents AD (Sui, Tang)

14th–19th cents AD (Ming-Qing)

750 km

0

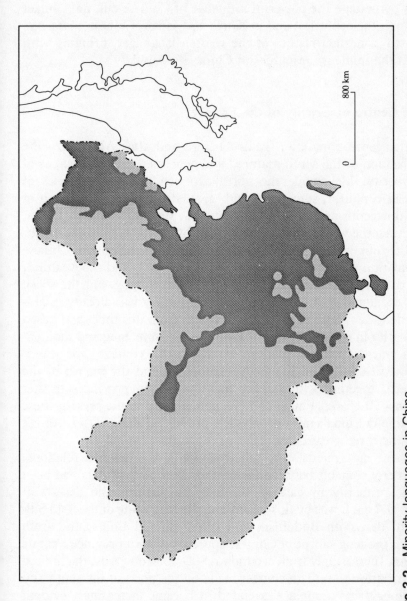

Figure 3.2 Minority languages in China

The dark shading indicates normal use of Chinese language (in its many dialects). Paler shading indicates general use of other languages, such as Mongol, Tibetan, Yi, and so forth – the original map shows 56 languages other than Chinese.

Source: Beijing Atlas 1984, 4

800 km

0

languages. Beyond the frontiers in Korea and Vietnam, Chinese surnames in the populations and Chinese vocabulary in the languages perpetuate the powerful influence of Chinese cultural contact through the centuries; and in Korea and Japan, Chinese characters are still a normal feature of the written languages, bringing with them the immense resources of Chinese vocabulary.

The Centre of Gravity of the Empire

In the generations after Tang – before and after AD 1000 – the civilization of the south matured and generated its own systems of prosperity. Meanwhile the north experienced further phases of nomad conquest, sometimes with great destruction. For much of the thirteenth and fourteenth centuries, first the north of China, and then the whole country, was ruled by Mongols from the north. By the foundation of the Ming dynasty in 1368 the bulk of Chinese population was already established in the south, and the wealth of the country still more so. Ming ruled from Beijing, but the social and economic centre of gravity of the country was already established at the mouth of the Yangzi. It was from this time, and based primarily in the south, that the empire began its long and momentous journey into the modern world. In this context, 'the south' means primarily the Jiangnan territory around the mouth of the Yangzi, together with Jiangsu and the coastal provinces further south – Zhejiang, Fujian and Guangdong. Up to the present these are among China's most progressive provinces, along with Liaoning and the three great cities.

These movements were reflected in China's external relations. Formerly contacts between China and other parts of Asia had been made primarily by caravan through the plateaus and deserts of central Asia; it was by this means and the long route of the 'Old Silk Road' that both Buddhism and Islam entered China, and from which the long salient of Chinese speech in Gansu province took its origin. Increasingly from around AD 900–1000 onwards, the limited contribution to foreign contact made by the ports of the south and south-east was gradually extended; it became increasingly evident that China had important neighbours in Japan and south-east Asia.

As time passed, contacts were made by sea by Western peoples which, however unwelcome, could not easily be resisted.

In terms of a geographical outlook which emphasizes location and locational structures primarily, the Chinese occupation of southern China and the migration of the country's centre of gravity to the south have been the most important changes in the whole historical geography of the Chinese subcontinent. In terms of an outlook which emphasizes resource and resource use, however, the tremendous growth of population in recent centuries is of equal or even greater importance, as the following section shows.

The Chinese Population Explosion

For many centuries, the population total of around 60 million does not appear to have been greatly exceeded; in fact the records of various subsequent censuses suggest that AD 2 was a high water-mark not easily reached again (Durand 1960; Loewe 1966). If a significantly higher population total (of more than 100 million) was reached around 1100 under Song, as some writers have suggested (Ho Ping-Ti 1959), it was not maintained through the warfare accompanying the subsequent Mongol invasions. The standard population figure for 1393, under Ming in the first generation after the Mongols, is given as 65–80 million (Perkins 1969, 216).

Where the total size of the empire had virtually doubled between Han and Ming, the failure of the population to grow commensurately might be thought surprising. It must presumably be attributed to the various positive checks to which population growth was subject, mainly warfare and disease. By the same token, the factors which permitted the subsequent formidable increase, where traditional conditions experienced little or no relaxation until the eighteenth century, are not really apparent. Moreover, there are exceptional difficulties in the use of census materials for the centuries when population growth was beginning.

Around 1550, however, the population probably reached 100 million again; around 1750 at least 200 million; around 1850, 400 million (Perkins 1969, 216). By that time the total population was so large that subsequent increase was bound to be immense: by

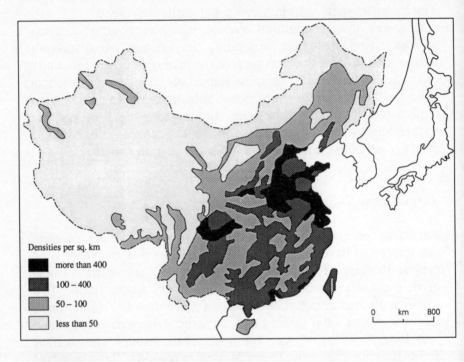

Densities per sq. km

more than 400

100 – 400

50 – 100

less than 50

0   km   800

Figure 3.3 Density of rural population
In China, most people are rural-based and density of population is
primarily an outcome of levels of rural production. In fact levels of
rural production on the local scale are higher in southern China than
northern, but southern China is rich in mountain masses with little
production or population, and on the regional scale represented by
the map populations are most dense (more than 400 per sq. km) in
Jiangnan at the mouth of the Yangzi, and further north in the wholly
arable countrysides of the north China plain. Outliers occur in
Sichuan, on the south and south-east coasts, and on the Wei River
in Shaanxi. Densities between 100 and 400 per sq. km generally
represent continuous settlement, for instance elsewhere in the north
China plain, the north-east and the middle Yangzi plains. Densities of
50–100 per sq. km generally represent intermittent settlement in
ridge-and-valley topography, especially in the south, or dry loess
conditions in the north. Densities below 50 per sq. km represent
scattered settlement or none, in mountain, steppe, high plateau or
desert conditions.
Source: Based on Beijing Atlas 1984, 3

1950 the figure must have been close to 600 million. By 2000 the total cannot well be less than 1200 million, and is likely to approach 1300 million.

This is the most dramatic expansion of a very large population that the world has ever seen. In common with other expansions of Asian populations in recent centuries, the increased population has been maintained, and is still maintained, almost totally from the soil of the country. Asian population growth, unlike that of the Europeans in the nineteenth century, has not been maintained by conquest and exploitation of distant continents.

In most of the Third World, rapid modern population increase awaited the introduction of peace and sanitation under the colonial empires in the nineteenth century. It is of some importance that in China the phase of rapid increase began at least a century earlier, and probably as early as the fifteenth century – and without the benefit of any but traditional sanitation systems. Traditional sanitation seems to have had one very important feature which may or may not have tended to limit diseases that can be spread by poor practice in this respect – that is the habit of collecting human waste, in both countryside and town, for use subsequently in agriculture. Most farm fertilizer in traditional times came from people or from pigs, composted with soil and vegetable matter.

The contemporary map of population density (figure 3.3) is the outcome of these dramatic increases.

The Expansion of Food Production

In these conditions, needless to say, Chinese farm outputs increased dramatically. Dwight H. Perkins (1969), the most distinguished writer in this field, shows that population increased between 1393 and 1957 by a factor of between seven and nine, and argues that food output was raised to cope with this increase by expanding the cultivated area and raising yields per unit area, in broadly equal degrees.

The cultivated area was extended by taking in hillside land (often using terracing); by greatly increased double cropping (for instance using a winter crop of wheat in fields which grew rice in summer, rather than fallowing); by extension of irrigation, so that riceland could be extended; by local reclamations from lakes or the sea; by ploughing up difficult areas such as those in the dry Great Wall frontier zone; and later by the extension of cultivation in the north-east (Manchuria), from which Chinese farmers were excluded by the Manchu aristocracy until late in the nineteenth century.

An increase of yields was managed during these centuries by improved varieties of crops, by more irrigation, by better fertilizer supplies, which came partly from the increased human and pig populations, and by more labour and more meticulous labour practices on the land. By the twentieth century, scientific plant nutrition and plant breeding were beginning to be understood; but the need for various kinds of fertilizer and the identification and differing properties of crop varieties were already understood in traditional times – certainly as early as Song.

One most important non-Chinese factor intervened in these developments, and that was the spread of the American crops to east Asia after 1500 (Ho Ping-Ti 1959). The American crops are common potatoes and sweet potatoes, maize and peanuts, together with tomatoes, peppers, chillis and tobacco. Those of most importance in the present context are the first four. All are tolerant of indifferent or poor soil conditions and all yield heavily. All were important in both phases of the historic expansion of Chinese farm possibilities. Common potatoes were valuable in encouraging cultivation in cool areas like the mountains of the south; sweet potatoes and maize in offering higher yields than millets in summer in the north China plain. Maize requires reliable rainfall but is otherwise very adaptable; its modern distribution is mainly in the long 'frontier' of indifferent farm conditions, extending from the north-east to Yunnan, which lies at the edge of the high plateaus.

According to figures assembled by Perkins (1969, 16), cultivated land per person under Han was 1.4 hectares; under Ming around 1400, 0.8 hectares; in 1957, 0.4 hectares. Needless to say, localities did and do display marked differences within these averages. Nutrition in terms of basic foodstuffs was probably not very different in the 1950s from the 1390s. An absolute rise in the extent of cultivated land had been more than compensated for by the much greater rise in the population, and land per person was much reduced. This in turn was compensated for by the rises in yield per unit area which have been outlined. In contemporary China, the same ambiguous relationships between improvement and population growth exist, in present conditions where the cultivated land area is now difficult to increase and is tending to fall. There is still substantial scope for technical improvements in Chinese farming (for instance, by comparison with Japanese), but population increase is still capable of taking up all the surpluses created.

The Mature Empire

Two stereotypes, both unrealistic, dominate Western thinking about the historic empire. One, derived from the missionaries in the nineteenth century and up to 1949, is of traditional China as ungoverned and ungovernable: lawless, chaotic, filthy, intolerable in its poverty, desperate. The other, derived from older experience less close to the people, is of an unchanging society ruled by a mindless bureaucracy and immemorial custom. Both found China heathen, hostile and full of unpredictables. In the present generation, with much better information and much better understanding of Chinese reality, it is easy to see how these two stereotypes came to be set up, and how inadequate they are. The nineteenth century was a time of crisis in China, for reasons which will be outlined in the next section; realistically, the chaos of that century is less representative of China's total experience than the cautious, highly controlled orderliness of the present. Nor, as has already been shown, was China without a course of development through the centuries; nor was the bureaucracy mindless, though its behaviour was strongly tied to precedent.

In the present context the later empire is more relevant than the earlier. It may help to review the geography of the empire, and to some extent the imperial system, around 1800. From 1644 to 1911 the empire was ruled by the Qing dynasty, of northern Manchu origin.

The population of the empire in 1800 was around 300 million – more rather than less – and was increasing at a rate approaching 2 million a year (Durand 1960, 247-9). Contact with the West was still at a rudimentary stage; the famous letter of 1793 from the Qianlong emperor to the British King George III outlines the status of China as a 'universal empire' and foreign rulers as tributaries – 'There is nothing we lack' (Teng and Fairbank 1963, 18–19). This was still nearly thirty years before the systematic exploitation of the opium trade by the British East India Company based in Bengal, and fifty before the Opium War of 1839–41. In fact the China of 1800 was rich in exports – tea, silk, porcelain, fine cottons, all keenly sought after in the West. All were produced in the south-eastern coastal or near-coastal provinces, Guangdong, Fujian, Zhejiang, Jiangsu and (for porcelain) Jiangxi. All depended upon the generosity of the environments in these provinces, together with craft skills among workers and business competence among their employers – the same group of assets as now promote business in these same provinces.

The Qing government resisted the import of opium because of its narcotic effects, but in fact opium imports turned over the balance of payments from positive to negative in the nineteenth century. The medium of international trade in east Asia was silver, brought by Western traders ultimately from South America. Silver became the standard for commercial transactions inside China, but the Chinese government never monetized silver, and currency was always scarce. Indeed, 'in a contented society with limited business activities, money was considered to be a necessary evil' (Yang 1952, 9). During the seventeenth and eighteenth centuries trade maintained a steady inflow of foreign silver, but a heavy outflow in the nineteenth.

The empire was full of activity – trade, land colonization, industry, agricultural expansion, military pressure on the frontiers, cultural enterprises. Inland provinces like Hunan were developed to provide grain supplies for the cities. In Jiangxi the porcelain businesses generated widespread trade in grain for the workers and timber for the kilns. Trade demanded transport, which was typically by water; the magnificent Yangzi system carried most of all. Colonization continued in the valleys of the south, on the semi-desert frontier in the north, and on difficult soils in the north China plain. In parts of the south-west, land could be had for settling well into the nineteenth century. Improved agriculture benefited from the growth of population and the increase of labour on the land, with progressive phases of intensification – poultry, pigs, fish, silkworms and cottage industry. Cultural enterprises such as dictionaries and anthologies confirmed the educated classes in the completeness and necessity of traditional cultural forms. Successful military adventures in central Asia in the seventeenth and eighteenth centuries are the foundation of the modern claim to Tibet, Xinjiang and other central Asian provinces; but in 1689 the empire made its first treaty with the Russian empire in its march across northern Asia.

The Qing government took little account of the growth of population during its centuries of rule, but government was nevertheless altered by it. Under the empire as now, the essential building block of central power in the regions was the county. Under Qing the population of a typical county doubled and doubled again. Inevitably this made for a weakening of the authority of the county magistrate who represented the Emperor, and by the same token the strengthening of other kinds of power – rich merchants, big landowners, syndicates of business-men and so forth. To this extent the weakness of legitimate authority and its unpredictable performance, which upset nineteenth-century travellers, were already taking shape in the preceding century. Similarly taxation, especially of urban business, lagged far behind its potential, due to official complacency and reliance on precedent.

Some Chinese and Western writers have envisaged various movements towards capitalism in the China of these centuries, pointing to the growth of big-scale commercial enterprises, the increasing division of labour, the control of production by mer-chant interests and so forth. It will be worth looking a little more closely at two kinds of locality and business in which this idea has been produced. Each is also important in its own right.

Settlement of New Land

Some indication has already been given of the increased scale and long course of development of the historic migration of Chinese families from crowded areas into under-exploited new land. For recent centuries, a good deal of detail exists about the reality of this experience, especially for one of the most important areas at that time, the mountains of the Qinling in central China between the

Yangzi and Yellow River basins, on the borders of Sichuan, Hubei and Shaanxi provinces. Review of some of this detail suggests the dimensions of the long colonization movement, at least in part.

In the Qinling, the phase of most dramatic development was from mid-eighteenth century to mid-nineteenth century (Fang Xing 1979; Xiao Zhenghong 1988). At that time the Qinling, much of it high mountain territory cut by deep gorges, was very thinly settled, with little agriculture.

In the earlier stages of colonization, families from other areas – displaced in local warfare or otherwise – moved to the lower mountain slopes, up to elevations around 1400 m, growing mainly maize and millets. Most of these people came from provinces to the south – Sichuan, Hubei, Anhui. In later stages, people took up land at higher levels, up to 4000 m. Later, too, people came from further afield, including Hunan and Jiangxi – extensive migrations, though not as long or as foreign as those to south-east Asia which were beginning to interest people in Guangdong and Fujian. In due course, recent migrants amounted to 80 or 90 per cent of local populations. In some counties, population densities rose between the late seventeenth and early nineteenth centuries from around 1 person per sq. km to more than 100, and from figures of much less than 1 to around 50 per sq. km. The people were 'free like crows'; when taxes and rents were proposed they resisted, and also argued that it was the people's labour which had created the value of the land in what had been wilderness.

Increasingly more varied agriculture was introduced. Migrants from the south identified suitable sites in valley bottoms and built paddy fields; and these people also brought up-to-date techniques in management of rice crops. Double cropping (using a winter wheat crop on the paddy fields, as on the Yangzi) was started. More labour, a much wider range of crop varieties and better practice raised yields. Potatoes, introduced later than maize, proved valuable at high elevations and were prized for their heavy yields; in many areas they were second only to maize as staple crops.

Economic crops gradually entered the farm systems in the mountains, especially in the valleys – tobacco was the first and most important, followed by cotton, ginger and peanuts. These commercial crops do not appear to have been sold outside the mountains, but of course the growing local populations would demand them, including peanuts for cooking oil. Tea, an old crop in the less wild eastern section of the mountains, was still primarily a commercial crop, and the same was true of crops from both wild and planted oil, lacquer and walnut trees.

The settlement and exploitation process also had an important industrial component. Land clearance produced timber and bamboo which (if transport was adequate) could be sold downstream. Timber also fired furnaces for the smelting of local iron ore. Paper works were based on local bamboo and other forest materials. Capital for workshops often came from local merchants with cash in hand. Enterprises could be substantial: around 100 men at a paper mill; in foundries, ten or a dozen men to a furnace, and six or seven furnaces to an enterprise; some dozens of workers at timber yards. Because of enterprises of this sort, the Qinling is one of the areas which some Chinese writers have proposed as producing 'sprouts of capitalism' in landscapes which, according to communist party orthodoxy, remained 'semi-colonial' and 'semi-feudal'.

The decay of this settler and land-clearance economy was not long delayed, and it resulted in part, predictably, from over-exploitation, in part from civil disorder. On lower slopes, by the later nineteenth century, there had been up to a century of maize cultivation with only limited fertilizer; yields began to fall drastically. Potato crops began to suffer seriously from disease. Where gangs of rebels formed, land went uncultivated. Yields were unstable; the price of food began to rise, doubling in the nineteenth century, owing to the increasing population and static or declining supplies. Industry fell into decay, partly because of an increasing shortage of accessible timber (upon which everything depended), partly because agriculture could no longer support it. Prosperity ebbed away and the area went back to subsistence with incidental commercial outputs – but of course with a greatly increased population.

This well-documented settlement story may serve as a model for the local experience of other settlement enterprises in traditional times in China, and perhaps in broad terms for the historic settlement of the south valley-by-valley and locality-by-locality – though in traditional times there was surely much more use of fire in the dry winter as a means of land clearance: 'fire tills, water weeds' as one ancient text has it. In Taiwan, where there has been much settlement during the twentieth century, peasants readily speak of recent woodland settlements, generally based on existing villages, where people began perhaps sixty years ago with forest exploitation but later took up agriculture, both rice and dry-field vegetable crops. This kind of local growth has been as fundamental to the creation of modern China as the high civilization of the empire or Maoist revolutionism; and as in other communities throughout the world, local opportunities and constraints cannot be separated from the detail of local, social and economic history.

The Jiangnan Industrial Economy

The settlement experience of the Qinling may serve as a prototype for a very broad and characteristic range of settlement experiences in China. The Jiangnan industrial economy was of a very different kind – far from typical, but very important and distinguished; arguably the most highly developed industrial economy in human history before the introduction of machine industrialization in the north of England in the eighteenth century. In Jiangnan too there were sprouts of capitalism, but capitalism of a very different kind from that proposed for the Qinling (Nishijima 1984). The Jiangnan complex was a large, advanced and very highly organized agro-industrial organization, which developed from the fifteenth century onwards, and came to its most elaborate state in the eighteenth and nineteenth centuries. In this environmentally favoured region, with a dense and growing population and advanced agriculture, a large cotton industry emerged which produced consumer textile goods for the mass market. The industry was in part urban, in small-to-medium workshops, but it was primarily rooted in part-time and women's work among farm families; and it displayed increasing division of labour. It was organized by merchants who sold yarns and bought finished cotton cloth. Local agriculture, normally within the same families, provided food. Local agriculture also furnished much of the cotton used, leading farmers increasingly into commercial farming. Families remained independent units of production, but tied to textile manufacturing as income supplement; inevitably the peasant families with their low costs and long hours of work were victims of a buyer market in this trade. Meanwhile population in the countryside continued to grow.

On the Pennine flanks in the north of England, in the same period as the full development of the Jiangnan system, a cottage industry system in some ways parallel to that of Jiangnan, but based originally on wool, led through organization and machinery to machine industrialism and the textile side of the Industrial Revolution (Deane 1979, chapter 6). No such development took place in China. This was not primarily through lack of knowledge or understanding of the possibilities of machinery (Elvin 1972). What seems to have been decisive was the lack of any incentive to develop machinery whose main purpose was to save labour. In China the cost of part-time rural labour was very low, and probably falling. Underlying the cheapness of labour was the crowding of a still increasing population in Jiangnan. In these conditions, the sprouts of capitalism received little nourishment and inevitably made no growth. But this regional economy was strong enough in its own terms to display marked resistance to competition from machine textiles both foreign and, later, domestic.

Origins and Foundations of Contemporary China

This chapter has concentrated on the origins and foundations of the Chinese community, and these have necessarily involved examination over many centuries. There is room in addition for some words of interpretation of China's more recent social and economic experience, particularly of the century before 'liberation' in 1949.

Chinese civilization has displayed great continuity, and its people have been, in recent centuries, extraordinarily numerous. Protected from change by an imperial system ruled by precedent, and by the people's own sense of the great size and individuality of China, the community both at the top and in its vast bulk remained ignorant of, and indifferent to, the immense changes elsewhere in the world during the eighteenth and nineteenth centuries. Upstart and distant foreigners cut little ice at the Chinese court when they turned up as embassies. Christian missions, after an auspicious beginning, were thought of as hostile and destructive, although at the same time tempting to those – mainly bourgeois – who were impressed by Western education and medicine. Western knowledge and understanding, in science and social science, were almost impossible to fit into the ordinary intellectual framework of Chinese thinking. China had been too inward-looking and too self-sufficient for too long; and the world had changed in fundamental ways which were bound to affect China, but about which even the best brains in China knew nothing. In the last decades of the nineteenth century, when Japan reconstructed her civilization with an economic and intellectual Western infrastructure, China fell into advanced decay: neither the dynasty nor the bureaucracy, neither the new economic forces nor the Westernizers, were able to manipulate this ancient and complex organization into constructive relationships with the world's new forces. Intellectual and economic disorientation both contributed to the long and disastrous Taiping civil war of 1850–78, which ruined the advanced provinces on the Yangzi. When the Qing empire was finally overthrown in 1911, the change was much less a matter of substance than of form, and the new Republic inevitably inherited all the old confusions and inadequacies.

This was the environment into which the Chinese communist party was born in 1920. It was one in which economic pressures, particularly the extreme poverty of the most crowded areas, subsisting within a technological framework not very different from that of three hundred years earlier, compounded the bafflement of an intellectual class whose education contained almost nothing which could be adapted to the understanding of the twentieth century. It was Mao Zedong, writing in the decades which followed, who provided Chinese intellectuals with a convincing diagnostic rationale of China's history and social conditions, based upon Marxist and Leninist insights, and stressing the role of Western and Japanese imperialism. It was Mao also who proposed a course of revolutionary action and open warfare which related to this rationalization. It was Mao's thinking in the 1940s, with its fusion of Chinese traditional outlooks with Marxist theory, which set the communist party upon the road to government. Regrettably, but in common with many others, Marxists in China turned out to be better at diagnosis than treatment; in Mao's time some problems of the preceding century, such as industrialization and security, were solved, but others, such as population growth, education and foreign contact, were compounded or simply ignored. By the time of Mao's death in 1976 the Maoist system had nothing more to offer.

FOUR

Structures and Determinants
in the People's Republic

Policy since 1949

The People's Republic of China was proclaimed in October 1949. Since that time the country has been ruled by the communist party in its various phases, and politics and policy discussion have been conducted almost entirely within the framework of that party. Essentially, the communist party has grafted its own outlooks and personnel upon the old official system: contemporary China is governed by officials who are perceived as all-powerful, as in traditional times, but these individuals are now members of the communist party and subject to its disciplines. However, not all thinking in the party, even in the leadership, has been harmonious or consistent; factions have come and gone, important individuals (particularly Mao Zedong) have exercised exceptional power and mainstream policy has undergone marked transformations. As a result, government and policy in modern China have showed some of the weaknesses characteristic of traditional times, especially unaccountability of the governing class and year-to-year instability of policy. Inevitably, some other traditional problems of Chinese management have also reappeared under the People's Republic, particularly those of the management of so large an area, with marked regional differentiation, and of so immense and varied a human population. Some traditional techniques of management have also reappeared: powerful social control fortified by mutual supervision among the people, and preoccupation with the need for order, if necessary at the expense of law.

Policy has been markedly unstable since 1949. Some of its various phases are summarized in table 4.1. Up to his death in 1976, these phases related primarily to the activities of Mao Zedong, who on several occasions overrode the responsible committees to turn the development of China in directions which he personally favoured – in 1956, 1958 and 1966 most dramatically. These interventions were of a 'leftist' character, proposing developments which should be 'big and public' – that is, large in local scale and collectivist in character, with marked hostility to individual property, business and management. In 1962 and after 1971 the pressures for maintenance of these Maoist systems weakened, but the Maoist inheritance was not decisively rejected in the ruling committees until 1978; even up to the present there are occasional signs of official interest in Maoist outlooks, especially when 'bourgeois liberalism' seems to threaten the communist party's control.

The definitive post-Mao reforms were set on foot in late 1978, two years after Mao's death. They and their results are the main

Table 4.1   Phases in Chinese official policy since 1949

1949–52	Period of rescue and initial reconstruction. Land reform.
1952–57	First Five-Year Plan on Soviet lines. Nationalization of most big business. Rural collectivization (from 1956).
1958–62	Unsuccessful 'great leap forward' and its aftermath. Departure from Soviet precedent. Collectivization of small businesses. Institution of rural 'communes'.
1962–65	Return to conventional 'economicist' and planning systems, but retention of most collectivist structures.
1966–70	'Great proletarian cultural revolution'. Universal adoption of 'leftist' policies.
1971–78	Slow retreat of 'leftist' policies and return to more conventional systems. Death of Mao Zedong, 1976.
1979–84	New beginnings. Return to family farming; reconstruction of commerce; 'open' development; cumulative return to conventional systems.
1985–88	Further liberalization – commerce, migration, foreign contracts, local discretion in management; but rural system tightened.
1989 to present	Rejection of political change; reaction against 'bourgeois liberalism'. Retrenchment due to overheating and inflation in the economy.

determinants of present development. They were much clearer and more persuasive for the countryside than for parts of the economy closer to the communist party, such as industry and the cities; indeed up to the present, with all their faults, the reforms in the countryside are at a different level of practicality and consistency from those in any other field. It is important to realize, however, that the reforms were essentially economic rather than political. No policies were adopted that were intended to reduce the role of the communist party in China, and none was seriously proposed except, as time went on, by dissidents. The official Chinese approach to the need for widened economic and social experience proposes continued management by the communist party as a self-evident prerequisite. 'Socialism Chinese style' is the objective.

There have been two changes of pace, and to some degree direction, in the reform movements since 1978. The first, associated with the name of the then Premier Zhao Ziyang, involved a range of liberalization and reorganization policies introduced in 1984 and 1985 – more freedom for business including the township enterprises, more personal freedom (vitally, to migrate), the redefinition of towns, the linking of rural counties with cities, the extension of rural land contracts. Taken together, these changes represented the adoption of market orientation in the economy. The second was the credit squeeze and deflation of 1989 and 1990, the result of overheating and inflation in the economy. The Tiananmen Square disaster of 4 June 1989 occurred during this phase, and has to some extent been treated in China as an aspect of necessary economic retrenchment.

It is clear that the communist party has not been idle during its forty-odd years of tenure of power. Policy has varied a great deal, not always rationally. Inevitably, there has been a price to pay for this history of experimentation and instability. The most important part of this price has been lack of continuity in local development, especially in the countryside and in trade. Another part of the price is decrease in social trust between the communist party (originally perceived as clean-handed and rational reformers) and the population, together with cynicism about official motivation and fatigue with social engineering and, indeed, socialist excesses. There is also a gigantic backlog of grassroots need for development, especially in the cities – of housing, urban infrastructure, transport networks, marketing and trading links, and so forth. This is because Maoist social engineering rarely had time or funds for social construction or modernization apart from state industry; and because as a result Chinese society inevitably slipped back by comparison with others in east Asia. The Maoists had little idea, in the 1960s and 1970s, how much constructive change was taking place elsewhere in Asia. Two kinds of inheritance from the period 1949–78 are particularly important – population and investment. These are both reviewed in sections that follow.

Population and Population Policy

The population of China passed the one billion mark in 1981 and the 1100 million mark in 1989. The figure for 1989 was almost exactly double the 1949 total of 542 million. In these 41 years population growth averaged around 14 million per year, or around 2.5 per cent. However, the average does not represent the real experience of the community, as figure 4.1 shows. The dramatic reversal of the main trends of all three functions in 1958–61, and the subsequent equally dramatic recovery of the birth and growth rates, was the outcome of the turmoil and final abandonment of the 'great leap forward' of those years. As figure 4.1 shows, the rise in the birth rate after 1961 more than compensated for the losses in 1957–61. The most rapid growth took place partly in the phase of political relaxation after 1962, partly in the Maoist 'cultural revolution' after 1966; growth rates were around 2.0 per cent from 1962 to 1973, mainly due to birth rates above 3 per cent. This has given China an immense population now in the young-adult bracket: some 250 million people, around 23 per cent of China's total, were between 18 and 27 years of age in 1990 (table 4.2 and figure 4.1). Attempts were made to revive family planning and later marriage in the 1970s, and in 1980 the policy of the one-child family was officially adopted, with rewards for compliance and penalties such as loss of benefits for transgression. This policy has understandably not been popular, mainly because of couples' desires to have a son at all costs, and a variety of relaxations have been permitted. Relaxations of principle exist among the minority peoples and in rural areas in certain provinces. Relaxations of practice occur in

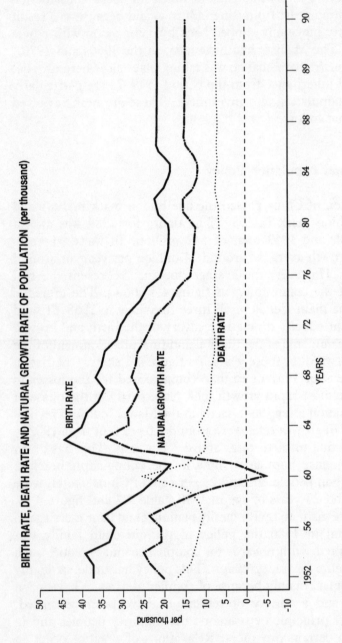

BIRTH RATE, DEATH RATE AND NATURAL GROWTH RATE OF POPULATION (per thousand)

Figure 4.1 Birth rate, death rate and natural growth rate of population
The diagram shows the dramatic fall in birth rate and rise in death rate of 1957–64, the phase of the 'great leap forward' and subsequent recovery. It also shows the long phase (1962–75) of birth rates above 23 per thousand, which has generated the present abundance of young adults in China.
Source: Redrawn and extended from SYC 1985, 187

Table 4.2   Population of China, 1949 to present (year-end)

	Population (millions)	Birth rate per thousand	Natural growth rate per thousand
1949	542	36	16
1952	574	37	20
1957	647	34	23
1962	673	37	27
1965	725	38	28
1970	830	33	26
1975	924	23	16
1979	975	18	12
1985	1050	18	11
1987	1081	21	14
1988	1096	21	14
1989	1112	21	16
1990	1143[a]	21	14
1991	1158	20	13

[a] The apparent exceptional growth in 1989–90 is the result of correction of the population total by the Census of 1990 (*People's Daily* 31 October 1990, 1).
Sources: *CSY* 1990, 81, 82. For 1991, *People's Daily* 29 February 1992, 2

rural areas where the first child is a girl, and in many family situations where registration of a birth may be avoided (Jowett 1989). Success of the policy appears to have been generally low throughout southern China, except perhaps Sichuan; its principal successes have been in the three great cities, among urban populations everywhere (strengthened no doubt by difficulties in finding accommodation), in the north-east and in Shandong and Jiangsu – all advanced areas with substantial urbanisation (Jowett 1989, 2, 75).

It was originally hoped that the one-child family policy would keep the population total below 1200 million at the end of the twentieth century, and that the population would start to fall slowly in the twenty-first century. However, the policy has achieved only limited success, especially among the rural majority; birth rates up to 1984 remained below 20 per thousand, but since 1985 have risen above that level. Fertility rates are now of the order of 2.3 to 2.4. All the relevant indicators are now stable, but as figure 4.1 shows, the difficulty is that they have stabilized at levels which are too high. The Census of 1990 offered little comfort.

For the future, figure 4.2 suggests four possibilities based upon four supposed levels of fertility (Hu Angang 1989). This reasoning insists upon a fertility rate below 2.2 for the foreseeable future, if population is to peak at a bearable figure of about 1500 million, around the year 2035. Higher rates set no limit upon potential growth. Particular problems are created in the present phase by the young-adult population 'bulge', since these people are now nearing

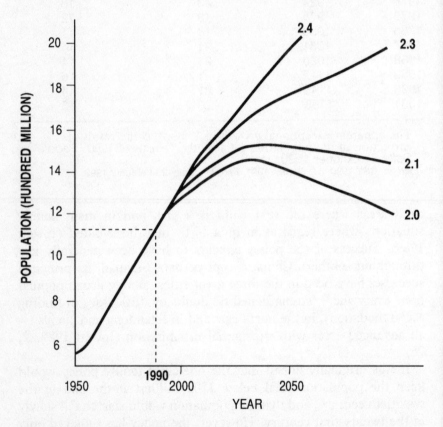

Figure 4.2 Projections of Chinese population
Projections according to various fertility rates. Sustained rates above 2.2 mean continued growth; below 2.2, eventual slow fall. Fertility in China is now running at 2.3–2.4.
Source: Hu Angang 1989, 57, 59

the age of marriage and child-bearing, and are also swelling the Chinese labour force by around 15 million per year, or 3 per cent – in a decade, 150 million or 30 per cent. Most of this additional labour is located in the countryside and so, of course, is most of the additional population. Inevitably, moreover, there will be particular difficulty with the burden of old people in the early 2000s.

Investment History

The inheritance of state investments made between 1949 and 1978 is also of critical importance for the China of the present. Table 4.3 gives a condensed view. State investments are cardinal in this field, because little investment of any kind has been made by other bodies – in left-wing phases, none at all. In table 4.3, heavy industry dominates throughout, with figures always above one-third of total investment and, in 1958–80, figures close to 50 per cent. One of the most remarkable and consistent features of China under the communist party has been the tenacity of the industrial bureaucracy in the defence and expansion of state industry, especially heavy industry. With continued investment in heavy industry goes necessarily the continued strengthening of the social classes and places which depend upon and benefit from it – not only the industrial bureaucracy itself, but the industrial working class and their bases, usually the big cities. In turn this distribution is reflected in the distribution of prosperity. In investment, light industry does not compete significantly. In terms of growth during these 40 years, the most marked development has been in the energy sector. This growth takes three main forms – coal, electricity (about one-half of it hydro-power) and oil, in which, due to the discoveries of the 1970s, China has become a medium-scale producer. Yet it is often pointed out today that investment in energy has not been sufficient. The investment experience of agriculture meanwhile has been revealing. Agriculture's proportion of Chinese investment has fallen from 7.1 per cent in the 1952–7 phase, through some unstable improvement in 1958–80, to a present low of 3.1 per cent. Meanwhile agriculture employs 60 per cent of the labour force. It is true that investment on the land is relatively

Table 4.3   Investment in capital construction by state units, by major sections of the economy (percentages, 1953–89)

	Agriculture	Light industry	Heavy industry	Energy	Transport and Communication	Othe
1953–57	7	6	36	12	15	24
1958–62	11	6	54	17	14	ni
1963–65	18	4	46	15	13	4
1966–70	11	4	51	16	15	5
1971–75	10	6	50	18	18	ni
1976–80	11	7	46	21	13	2
1981–85	5	7	39	20	13	16
1986–89	3	7	43	26	13	8

Source: *CSY* 1990, 155. Figures for 1986–89 have been recalculated from annual data. 'Others' a category is derived from figures in the original tables.

cheap and seldom comes in large units; but the systematic starving of the farm economy of investment (and hence innovation) has been unwise. Much was written in China, at the time of the reformulation of policy around 1980, about the 'lop-sided' development of the heavy industrial economy at the expense of light industry, agriculture and consumer standards; but the evidence of these figures is that the powerful industrial vested interests have continued undisturbed on their way. 'Others' in table 4.3 represents non-productive investments in fields such as schools and housing. Investment in these fields sank to starvation levels in most phases between 1958 and 1980, which is one reason for the immense backlog of low standards and obsolescence which disfigures the Chinese community.

In principle China has maintained five-year planning, although continuity was interrupted in some phases by political movements. In principle also, five-year plans in a centrally planned economy can reasonably be expected to arbitrate among various kinds of investment – transport, industry, mining, agriculture, health, education and so forth. In fact, China's five-year plans have not been of this kind – realistically, they have been industrial plans first and foremost. This relates to the point made earlier about the tenacity of the industrial bureaucracy. Needless to say, concentration of investment on industrial development in the state sector for nearly two generations has created industrial problems of its own, which will be introduced in chapter 6 on industry. More generally, the accumulated force of state investment has made its principal contribution to Chinese regional differentiation, as will be shown in chapter 8.

Regional Structure

China is a very big country, and the essentials of its management cannot be separated from its regional diversities. To introduce them, it seems best to adopt Beijing's own formulation, which is represented in figure 4.3. This regional scheme was devised in 1986, explicitly in terms of development opportunities and priorities. It separates out 'East', 'Centre' and 'West' regions of the country (figure 4.3). In development terms, the East region is the most developed, with relatively advanced technology, a skilled workforce and high incomes, overseas connections and opportunities. The West region is the least developed, with many undeveloped resources but great need of new technical assistance. The Centre region falls between the other two in terms of its opportunities and needs, but is considered to be moving relatively rapidly towards East status.

Figures which relate to the three regions are given in table 4.4. The main point of this table is not the absolute size of the figures, but their relationships. In total outputs, the East scores almost four times the figure for the West, and in outputs per square kilometre nearly 16 times. Incomes per person, on the other hand, are much closer (only about double), and the same is true for incomes per unit of investment. For some indicators, the Centre region comes closer to one or other of the West and East, but for others the Centre stands equally between the others. In part this is because of the very mixed character of the Centre region's territories (including the dense agrarian populations of Hunan and Hubei with the scattered fringe settlements of Inner Mongolia), in part because of the implications of each indicator. In general table 4.4 serves to illustrate the distinctions which the central authorities intended the regionalization to represent. However, these authorities cannot be happy with another point made by Chinese writers (Zhou Shulian

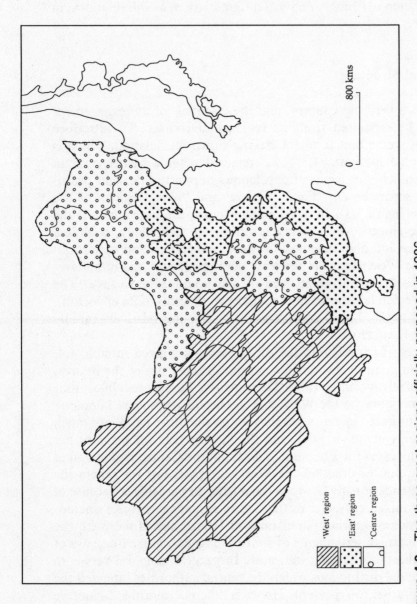

Figure 4.3 The three great regions officially proposed in 1986

The three regions are intended in part to recognize differences which exist on the face of China, in part as forming the basis for differing policy intentions.

Source: Dili zhishi (Geographical Knowledge) 1986 (7), 10

'West' region

'East' region

'Centre' region

0   800 kms

Table 4.4 The three major regions – various economic indicators (1985)

	East region	Centre region	West region
Area (thou. sq. km)	1294	2820	5414
Cultivated area (thou. sq. km)	32	42	24
Population (million	430	371	240
Total output			
(billion *yuan* at 1980 prices)	920	491	242
Industrial output			
(billion *yuan*, at 1980 prices)	566	245	113
Farm output			
(billion *yuan*, at 1980 prices)	126	106	59
Population density			
(persons per sq. km)	332	130	44
Outputs			
(thousand yuan per sq. km)	711	174	45
Income per person			
(yuan, at 1985 prices)	896	595	462
Income (1985, yuan) per			
thousand yuan of investment in 1952–85	10,600	7,100	5600
Railway density			
(km per thousand sq. km)	1150	850	240
Road density			
(km per thousand sq. km)	23,420	11,710	5610
Revenue surplus			
(billion yuan)	23	−5	−7

Source: Zhou Shulian and others 1990, 64–5

and others 1990, 73), that according to three general indicators (gross outputs, total incomes, and incomes per person) there has been a very broad and very consistent tendency, ever since 1952, for the differences between the three regions to widen, always to the advantage of the East region.

Obviously this regional scheme has weaknesses, both in using only three regional categories and in using only the provinces (realistically less than 30 in number) to define them. Differentiation is much more intensive than this, and reality much more complex. Sichuan lies, according to this scheme, in the West region. From the standpoint of its isolation this may be quite appropriate; but the most highly developed parts of Sichuan, around Chengdu and Chongqing, are more intensively developed than many inland parts of coastal provinces such as Guangdong, Fujian or Shandong. Jilin in the north-east is a prairie province with a strong industrial component; Hunan and Jiangxi, which share Centre status with Jilin, are old-fashioned paddy provinces. Yet all three have in common that they are important suppliers of commodity grain to the state.

Most of the provinces allocated to the West are primarily 'minority' units in that either the province itself has *zizhiqu* (autonomous region) status, like Xinjiang, or important parts of it have similar status at county level. In the case of Xinjiang, Tibet, Inner Mongolia and Ningxia, all in the north or west of the country, and all with indigenous populations of central Asian types, these units are large and historically cohesive, and (among other things) may wish sooner or later to follow the former Soviet republics in making their 'autonomous' status rather more real than at present. Inner Mongolia, however, is allocated to the Centre tier of provinces for regional purposes, while Sichuan, Yunnan and Guizhou, all provinces which are unquestionably parts of China in spite of being rich in autonomous (minority) counties, belong to the West. Guangxi, the only 'autonomous region' province in southern China, is also unquestionably Chinese, in spite of its official 'Zhuang' nationality status. Guangxi, no doubt as a coastal province, comes within the East regional grouping, in spite of its relatively low levels of development.

Finally, there is the important consideration of complexity. In China as in other countries prosperity (and usually development) occupies cells which are much too small to be defined by provincial boundaries, and which can be very varied in dynamics – from state industrial investment in southern Liaoning and the Wuhan countryside to overseas Chinese investment in the Pearl River delta, or historic intensification systems in southern Jiangsu or the central basin of Sichuan.

Official Management in China

In traditional times, the source of power in China was the emperor, and the management of power took place through civil servants whose authority was perceived, like the emperor's, as practically limitless. The characteristic geographical units of imperial power were the counties, usually numbering between 1500 and 2000. Powerful families, like the feudal gentry of Europe, Japan and India, could flourish locally, but except on the frontiers and at times of civil disturbance they had no authority to compare with that of the country magistrate; if such families wished to exercise official power, they had of necessity to seek official rank and appointment. There was no substantial authority except that which came from the emperor.

It is not too much to say that in contemporary China the communist party has adopted the same structure of authority. Policy (now of course much more pro-active than in traditional times) is decided upon in Beijing and made known to the officials (who are communist party members) through hierarchies of contact extending down to the 1919 counties and 162 county-level towns, still the most important local units. In modern times, of course, the media also operate as an important instrument for the dissemination of official policy; and so do communist party branches, which extend down to local small towns, and which perform some of the functions of local social management formerly exercised by important local families. Where Beijing permits or encourages variants of standard policy in particular provinces or regions (as in Guangdong province and the Special Economic Zones), provincial and even local officials may be able to exercise more independence, though always within limits. In effect local officials, particularly at county or equivalent urban level, are perceived as exercising the authority of the state itself, as in traditional times that of the emperor. But as in traditional times, their appointment depends on their trustworthiness as perceived within the hierarchy; and their behaviour is always potentially under scrutiny.

The administration of day-to-day life in China is performed through this bureaucratic hierarchy, which has developed relatively smoothly since the middle 1970s, after its violent shake-up in Mao Zedong's Cultural Revolution of 1966 and succeeding years. Individual bureaucrats are usually well-educated (bearing in mind the isolation, and in some phases obsessions, of Chinese education since 1949) and very intelligent, and the system they operate has some built-in leanings towards benevolence. At the same time, the system depends at every level upon an unself-conscious insistence on social control and, to promote it, mutual supervision among the people. For the officials, whose power is perceived as absolute, corruption is an inevitable temptation. In the present phase, corruption in which money changes hands is perhaps not as usual as might be expected, though it certainly exists at all levels. What is even more often indicated is: the use of official power for personal or family advantage, for instance where business contracts or land availability are concerned; nepotism towards relatives or friends or friends of relatives; habits of luxurious free meals and free travel; mutual back-scratching among officials; a powerful sense of authority and privilege with only one's seniors to answer to; an easy lifestyle with clean hands and an indoor job; social advantages for one's family; even such bonuses as trips abroad. None of these things is immune from sharp criticism from Beijing, and people of high rank convicted of corruption involving money are executed from time to time. However, a system which delegates apparently limitless power to all officials, down even to local policemen, is bound to perpetuate both the acceptance of authority and the possibility of its abuse. Elected assemblies exist in China at county level and above, but it is rare for them to challenge official management in any way – indeed they are also often the subjects of official management. Overriding official authority is a continuing problem throughout south and east Asia, even in countries like Japan and India which have much more lively representative institutions than China; but in China it is a problem of special prominence.

Bureaucracy and official control are also of importance in the higher reaches of policy. It is common that where functional problems are perceived within the state system, the solution adopted is essentially a bureaucratic one. Thus, in 1984 the problem of slow assimilation of rural counties to the new reformed schemes of development was tackled (it was believed) by the reconstruction of the administrative hierarchy, putting many of these counties under the 'leadership' of adjacent cities. When in the later 1980s the problem of ownership rights versus operational practice in state industry was raised, a series of official bureaus was established in order to monitor and guide development, so that both unrealistic intervention by the owners (state organs) and improper exploitation by the operators (management and workers) should be prevented (Luo Yuanming 1990). One common characteristic of directives from the communist party on any subject is the insertion of long lists of aims, many of them obviously unrealistic. Thus, acknowledging that much of China's present output of coal comes from township mines operated by peasants, Li Peng (then vice-premier) invited these peasants to go on to invest capital in agriculture and forestry, to improve grain outputs, to plant trees and grass and keep animals, run transport and processing industries, develop cement production, build power stations and start chemical industries based on coal (Xian broadcast 1985). Not in this case, but in many others, this kind of practice appears to result from systems of committees in which each member's pet idea finds a place in the report.

The Economic and Political Climate

During the later 1980s, a 'democracy movement' gathered steam in China, to meet with violent suppression in Beijing in 1989 in the 4 June disaster, following upon several weeks of demonstrations in Tiananmen Square. Its spearhead throughout was the student community, who were perhaps less concerned with 'democracy' than with practical liberalization. The ruling communist party committees were already faced with serious problems in the economy – overheating owing to over-rapid development, and inflation in double figures and rising, according to some reports, to 40 per cent. In 1988 and 1989 various measures were already being taken to suppress demand. The 4 June disaster stimulated widespread hostile Western reaction and withdrawal of economic support. The broad result of those two bodies of experience has been reaction by the central authorities against political or economic adventurism, a return to tighter official control where it had been loosened during the 1980s, and a retreat to hard-line communist party outlooks. Rejection of the communist parties of Eastern Europe and the Soviet Union by popular movements has also resulted in the hardening of party thinking and discipline in China; no political reformism is underway or is officially contemplated. At the same time the communist party remains committed to its broad economic policies of growth, strengthening of business linkages of all kinds, technical modernization and restructuring to the advantage of consumer need in agriculture, light industry and trade. It is often thought in the West that a liberalized economy cannot be separated from a liberal society; but in east Asia a number of advanced economies have flourished in recent decades in societies without free politics – in South Korea, Taiwan, even Hong Kong. The Chinese communist party aims for a parallel accommodation. The state's aim remains 'socialism with Chinese characteristics', and a 'planned commodity economy' (Wang Mengkui 1991). Even in Western countries, 'democracy' is often conditional upon various official management systems which may amount to 'electoral dictatorship'. In China the long tradition and recent habit of direct rule by unelected officials lends strength to any government, of whatever philosophical outlook, which produces reasonably clean and competent administration, security and an adequate degree of prosperity, even without the gloss of multi-party democracy.

FIVE

The Countryside – Rural Development and Rural Problems

Fundamentals

The Chinese peasantry may be considered the largest single social group on earth, and the Chinese countryside, by the same token, the earth's biggest single problem of rural development. It is a problem whose Herculean size has been matched by no less Herculean efforts of reorganization and reorientation since 1949; and although there has been much wasted effort and failure of various kinds, on the time-scale of decades which we are now able to use there has also been solid achievement, and in most communities a realistic fresh start, especially in the phase of relatively stable policy since 1979. Given that the population of the Chinese countryside is now of the order of double that of 1949, achievements in development have been a condition of survival, but at the same time achievements have been constantly at the mercy of rising consumption. In fact over most of China rural living standards have risen dramatically since 1949. It is not too much to say that the creation of security and a reasonable measure of prosperity in the countryside has been the greatest of all the many achievements of the People's Republic in these forty years, but it has led, of course, to a rising population with rising consumer expectations.

Environmental differences between the eastern and central Asian halves of Chinese territory have already been outlined in chapter 2, together with those between the northern and southern halves of eastern China. These environmental differences are the foundation of traditional differences in rural land use which remain funda-

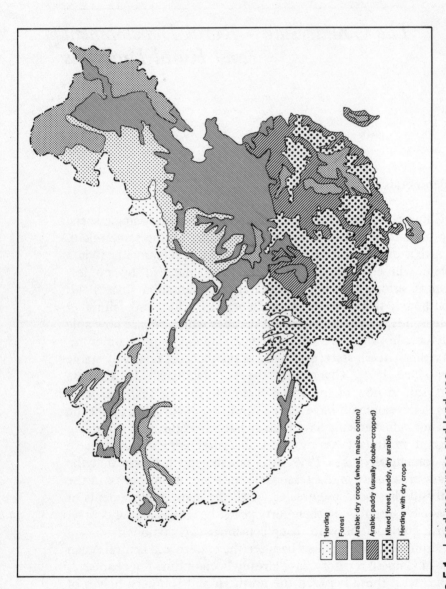

Figure 5.1 Land resources and land use

The map indicates the broad categories of regional land use in China – herding, forest, arable of various kinds. All relate directly to land resource and opportunity.

Legend:
- Herding
- Forest
- Arable: dry crops (wheat, maize, cotton)
- Arable: paddy (usually double-cropped)
- Mixed forest, paddy, dry arable
- Herding with dry crops

Source: Middle School Atlas 1978, 14 and 15 (reproduced by permission from Geography 1989, 74(4) 350)

mental up to the present. Northern China, with its loose, dry and calcareous soil, is dependent upon 'dry' grains like wheat, millet and maize; southern China, with its summer abundance of rain and sticky, non calcareous soils, is part of the east and south Asia rice system which extends from Japan to Sri Lanka.

Figure 5.1 indicates the most important features of rural land use in China. It separates out, first, the vast pastoral territory of central Asia, with its patches of mountain, forest and oasis agriculture, and its very extensive periphery in northern and north-east China, where agriculture is practicable in many localities but (due mainly to drought) by no means all. This vast area is occupied mainly by non-Han peoples akin to those of the republics of ex-Soviet Central Asia – Mongols and Turks. Here the majority of rural people are pastoralists; many are still nomadic. Figure 5.1 also shows major areas of forest, mainly in mountain areas of the south-east and centre, or parts of the north-east which are still peripheral to the main colonization movements; and it indicates very large areas in the south-west and south-east where remnant forest is varied by agriculture, which may be rice paddies (where land can be levelled and water supplies are available) or dry fields growing maize, upland rice or sweet potatoes. Much of the south-west is occupied by non-Han peoples who have their own systems of land use, sometimes involving shifting cultivation even up to the present.

The remaining areas are primarily arable, and as figure 5.1 shows, fall into two groups – north and south of the Qinling–Huai transition zone. South of this zone rice is the preferred crop wherever conditions are suitable; here the land is very often double cropped. North of the Qinling–Huai zone is arable land used for dry crops – the north China plain, the plains of the colder north-east and the plateau and valley country to the west, leading to the central Asia grasslands.

The basic distinctions in rural land-use and livelihood systems are not in any way the outcome of management under the People's Republic. They have evolved through many centuries, mainly in terms of the gradual establishment of farming systems and their gradual assimilation of surrounding forest and grassland, as has been shown in chapter 3. Only the north-east, which has developed mainly during the present century, is a major exception.

The pressure of cultivation upon the forest and grassland peripheries is now acknowledged to have reached the level of serious environmental weakening in recent years. But although the cutting of forest and opening of grassland are now limited, there are many other respects in which environments are under heavy pressure in rural China – cutting of bamboo for industrial use, cutting of timber for fuel, overgrazing or continued cultivation of grassland, stimulation of widespread soil erosion, and so forth.

The agricultural systems which developed in the long period of population growth under Ming and Qing were increasingly labour intensive – that is, they depended increasingly upon the heavy inputs of labour made possible by population growth. Abundant labour made possible the rapid turnover of crops in summer involved in double cropping, though at the cost of idleness at other times. Labour was also an essential input into the many systems of diversification which evolved, especially in advanced parts of the south such as Jiangnan, the area at the mouth of the Yangzi, and the Pearl River delta around Guangzhou in Guangdong: cotton, pigs, chickens, ducks, silkworms, vegetable gardening, fruit and the rural textile industry which has been outlined, and which worked in both cotton and silk. Most of these specializations depended at bottom on local outputs of food, whether for people or for animals. Labour could even be converted to capital, through investment in terracing, new water control systems, irrigation water and so forth.

The Maoists adopted essentially the same philosophy – of a densely populated, self-sufficient countryside in which abundant labour, its productivity now somewhat raised by modern scientific methods and possibly by mechanization and electrification, should produce its own food and where possible a surplus, and should go on to schemes of development, including various kinds of diversification (Mao Zedong 1956). That at least was the scheme of the 1950s; by the 1960s the Maoists had revised their thinking, and no longer wished to see any significant private or even collective rural production if it involved sale through a free market. Most rural markets were closed. When rural reform began to take hold in the early 1980s, a return to local marketing was one of the reforms most welcomed by the peasants. For some years, renewed progress with both intensification and diversification, including local industry, attracted warm praise in the official media (Leeming 1985a), and dense rural populations remained one of the starting points of the system.

In 1984 the communist party proposed a new concept for rural China – essentially a commercial countryside. 'Transformation from a self-sufficient or semi-self-sufficient economy to large-scale commodity production is a necessary process in developing the socialist economy of our country' (Chinese Communist Party 1984, K1). The household farming of the reforms should be retained and developed, but with fewer households each occupying more land. Households' contracts should be extended to at least fifteen years, and be variable to suit the farmers, with the collective's consent. Commerce should be developed afresh. Many more rural industries should be developed, and many more peasants encouraged to take jobs in them. Migration of peasants could be allowed in certain circumstances, provided that they were prepared to find their own supplies of food – a relaxation which rapidly expanded. The hiring of labour was still hedged about with conditions, but there were signs that the conditions could be helpful. Rural enterprises of all kinds, both collective and individual, were now said to have their proper functions, particularly the employment of surplus rural labour. At the same time, the communes and their subsidiary units were being dismantled. Along with this reform the state abandoned the institutions which gave the rural planning system the force of law. From 1984 onwards, the communist party set out to establish a commercial countryside with massive diversification, a smaller population due to migration, whether to local towns or beyond, much more specialist farm production from fewer and larger farms, and higher outputs due (it was expected) to economies of scale, more effective use of modern inputs, mechanisation where practicable and specialisms. Socialism was to be maintained in this system, it appeared, by two conditions, one formal and one informal. The formal condition was and is the important one – that rural land continues to be the property of the collective unit. Typically this is the village, though anomalies can arise, such as the claims of the local township to which a village 'belongs'. The informal condition is the continuing capacity of local officials to intervene in the activities of the people – thus, enterprises require licences, which are granted by local officials. Local officials may be corrupt, but the

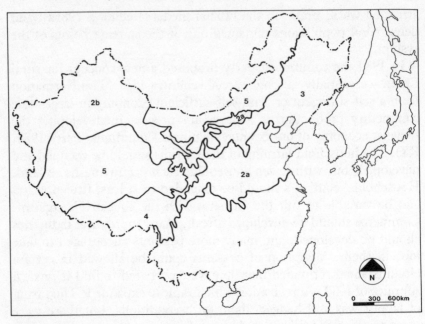

Figure 5.2   Crop association regions
This map separates regions on the basis of the characteristic staple
food crops. The key below summarizes the crop associations for
each region.
1   Agriculture based on rice: Typically two crops annually – rice with
     rice, rice with wheat, rice with maize or millet; sometimes other
     crops such as peanuts, soya beans, rapeseed, sugar-cane,
     vegetables. There are many substantial areas of mountain forest
     in this region.
2a   Agriculture based on wheat: Typically winter wheat combined
     with summer cotton or millet or tobacco or maize or soya beans
     or sweet potatoes. Two crops annually or three crops in two
     years (alternate winters uncropped).
2b   Agriculture based on wheat: Oasis agriculture in scattered
     locations; summer cropping only or three crops in two years.
     Alternative crops – maize sugar-beet, cotton. Most of this region
     is without agriculture and depends on pastoralism.
3   Agriculture based on maize or millet: Summer cropping only;
     alternative crops are wheat, soya beans, sugar-beet, millet, rice.
     Important parts of this region are forested.
4   Agriculture based on barley or oats: One crop only; alternatives,
     wheat or potatoes. Much of this region is too steep or rocky for
     cultivation.
5   Regions virtually without agriculture.
Source: Beijing Atlas 1984, 17

official system is alert to corruption, though not always active in its suppression. Officials may also have their own interests to serve, as will be shown.

In part, perhaps primarily, the new rural policies must have been a reaction to the problem of surplus labour. Surplus rural labour was a point of occasional discussion in the 1950s (Mao Zedong 1956), but it disappeared as a topic in the 20 years of Maoist ascendancy. When it reappeared in the media, in the late 1970s, it was being claimed that in many villages between 30 and 40 per cent of local labour was surplus to need.

It is the system of 1979–81, together with the further changes of 1985, which forms the basis of present rural development institutions in rural China. In so large and varied a system, where official policy can exert so much leverage, it is not surprising that on the one hand change can be very rapid, or that on the other policy may have unintended side-effects.

Cropping Systems

The two maps of cropping systems (figures 5.2 and 5.3), both taken from recent Chinese originals, illustrate two different aspects of rural China's great variety. Both are gross generalizations from much more detailed maps. Figure 5.2 is based on the staple food crops in each area, which everywhere are combined with other crops, some also staple, some industrial. The map identifies the southern region based on rice, two very different regions where wheat is the main food crop (the north China plain and its western extensions, and the scattered oasis agriculture of the north-west), one based on maize or millet (mainly the north-east), and lastly the limited agriculture of Tibet, based on barley or oats. The complex of interdigitating territories shown in the centre of the map is the outcome of highly marginal environments at the peripheries of central Asia. The key to the map names various alternative crops for each region, some staple, some industrial. In some cases these alternatives represent rotations from year to year, but where there is double cropping, rotation takes place within the year. In the north China plain (region 2a), wheat is generally grown as a winter crop harvested in May; it can normally be combined with a fast-growing

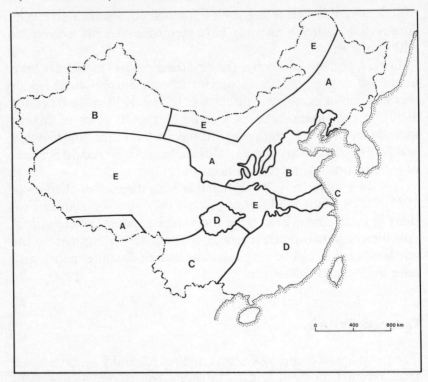

Figure 5.3　Cropping systems in China
This map is based on the identification of regions according to
frequency of cropping. These are as follows: (A) one crop annually;
(B) two crops annually or (more often) three crops in two years;
(C) two crops annually, usually including rice; (D) two rice crops
annually, sometimes plus a third crop; (E) virtually without
agriculture. In figure 5.3, the regions shown as 'virtually without
agriculture' differ considerably from those with the same description
in figure 5.2, apparently due to the differences in mapping criteria.
Source: Based on Hou Xueyu 1979 (this simplified version is reproduced, by
permission, from Leeming 1985a, 14)

crop such as millet, but if cotton is included in the system its long
growing season will preclude the use of another crop in the
following winter, and a green fertilizer may be grown. In the
north-east the winter is severe, and the crops named are alternatives
from year to year.

A second map (figure 5.3) gives an alternative perspective. This map has been constructed on the basis of frequency of cropping rather than kinds of crops. Double cropping is a basic technique in Chinese agriculture. By this means two crops can be taken in a single year; alternatively three crops may be taken in two years by a less intensive version. Double and semi-double cropping are possible in China, as in many other subtropical countries, because the winters are mild and dry; it should be recalled that the latitude of Beijing (40°N) is close to that of Lisbon or Philadelphia. In fact Chinese latitudes are rather cold in winter by global standards, and winter cropping is hardly possible north of Beijing.

Both maps distinguish the rice region of the south. Throughout this region, the staple of diet is boiled rice. Throughout also, double cropping is common and in advanced areas normal; many areas in the far south grow two crops of rice annually, between March and July and July and November. This kind of management is greatly aided by the usual practice of transplanting rice seedlings from a nursery bed. Others grow a summer rice crop against winter wheat; the latter occupies the ground from November to May. Rice, however, needs level fields which can be flooded. Particularly in mountainous areas like the south-west, summer rice is combined with other summer crops like sweet potatoes or maize, or with winter wheat.

The north China plain and its peripheries on the plateau to the west are also distinguished on both maps. Rice is difficult to grow here, not because the hot summer is inadequate but because soils are generally loose and calcareous, fertile but unretentive of water. Wheat (eaten as either steamed bread or noodles, both made with wheat flour) is the principal staple of the plain in modern times, though the old staple was millet porridge, and many country people still prefer it. Wheat in this region is normally grown through the winter, as further south. As a result, the fields are available for summer cropping – of sweet potatoes, maize, millet or cotton. Because wheat and cotton both demand a long half-year in the fields (wheat through the winter and cotton through the summer), double cropping with wheat and cotton is not practical in these latitudes, and even millet or maize are difficult to fit in between two winter wheat crops. For this reason semi-double cropping is often used, giving a winter wheat crop followed by summer sweet potatoes, followed by fallow until the next summer crop, which can be started early to allow time for a following winter crop. In central China (region C, along the Lower Yangzi, an area of extremely dense population), a very important and characteristic double-cropping rotation is used, of summer rice with winter wheat.

Broadly, the north-east is too cold in winter for winter wheat, and the same applies to the high plateaus along the Great Wall frontier in the north. This is the world of one crop per year, and at the same time of maize and millet staples rather than wheat. Maize and millet are both fast crops well adapted to a short, hot summer with more moisture than the north China plain; and maize in particular had strong official encouragement in Mao's time because of its heavy yield. Soils in much of the north-east are of good prairie or forest types; the problem (in spite of very moderate latitudes (50°N in Heilongjiang)) is the prolonged cold of winter. Tibet, too, due to elevation and consequent cold, is one-crop country.

In most of these schemes, green fertilizer (which can be fed to animals) is a common filler between substantive crops. Melons and vegetables are increasingly important as standards rise, and they can also be very profitable. Fruit is important in many places – stone fruits like apricots and peaches, the old-fashioned Chinese persimmons and jujubes, apples and pears. In the south these are supplemented by oranges (which are native to China), lichees and others. A few food crops, notably potatoes and sweet potatoes, are important, especially in poor areas, but do not figure prominently in the literature. Potatoes, like maize, are difficult to fit into Chinese diets and so not popular; but in the early 1980s some 8 per cent of food-crop acreage was cropped to potatoes of some kind. By far the largest part (72 per cent) of Chinese arable land is cropped to food staples; 'industrial' crops such as cotton and tobacco occupy 15 per cent, vegetables 4 per cent, green fertilizer 3 per cent (*CSY* 1990, 342–3).

Yields of Grain

In Chinese agriculture, where arable land is generally scarce, much depends upon local and regional crop yields. These are shown in figure 5.4, also taken from a Chinese original. Detail in the map dates from 1975, but provinces where there have been marked increases in yields between 1981 (the first for which there are published provincial figures) and 1989 are shown with a star.

The first point to note in interpreting this map is the predominance of high yields south of the Qinling–Huai zone. The reason for this is three-fold – in part the result of the predominance of rice in the south, with its typically high yields, in part the outcome of more favourable climatic conditions generally, in part the outcome of double cropping, common south of the Qinling–Huai zone but much less so north of it. Crude calculations of grain outputs per unit of cultivated area (there are no exact figures) typically suggest figures for southern provinces as a whole of more than 6 tons per hectare and, for some, more than 8 tons. Double cropping in southern China typically takes one of the two forms already outlined, both involving rice. In the Yangzi region summer rice (June to October) is usually teamed with winter wheat (November to May); but in the far south two rice crops are taken (February to June and July to November). These systems are traditional, but of course have benefited in terms of yields from modern inputs such as fertilizer and pesticide.

A second important feature of figure 5.4 relates to yields of less than 3 tons per hectare. These areas of low (and often unreliable) yield occupy a long and almost continuous belt of territory running north-east and south-west through the centre of China's territory. In effect this is the peripheral zone between Pacific China and central Asia, wider in the north-east and Inner Mongolia than in the south-west where it backs up against the Tibetan mountain ranges. Here, of course, double cropping is not possible, and physical conditions, of topography and (in the northern section) rainfall, are poor.

A third important feature is the several large areas in the north China plain, and southwards to the Yangzi, which also have low yields. These are due to a variety of physical features, especially poor soils, waterlogging or drought, and because they are usually areas of high population density they are the cause of particular anxiety.

China as a whole registered an increase in grain yields per unit area per crop of 16 per cent in the period 1982–7. Provinces in

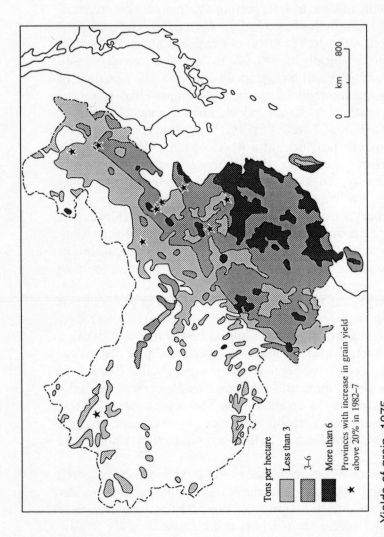

Figure 5.4 Yields of grain, 1975

The map shows yields in tons per hectare of cultivated land. These are clearly higher in the southern half of Pacific China than the northern; and in both regions there is a western fringe of lower yields, which abuts upon the central Asian massifs, dry in the north and precipitous in the south. Stars indicate the provinces with the most impressive increases in grain yield during the phase of rapid increases in the 1980s. All are among or peripheral to the northern provinces.

Tons per hectare

Less than 3

3–6

More than 6

★ Provinces with increase in grain yield above 20% in 1982–7

0   km   800

Source: Redrawn from Geographical Research Institute 1980, 121

which these increases were more than 20 per cent are marked with a star on figure 5.4. They are all in the north, which indicates both the long-term stability and relatively high levels of traditional technique in southern rice farming, and the marked broad increases in wheat yields (of around 25 per cent), gained in the same period. Increases in both Jilin and Heilongjiang in the north-east probably also reflect more economical and more intensive husbandry in these provinces, with their pioneer quality whose earlier extravagance has often come under criticism.

In China as a whole, grain is normally scarce. Rural populations who grow their own food are dense; at the same time the state needs to buy grain from the peasants to supply the cities, deficit areas, the army and so forth. Consequently the politics and economics of grain supply are matters of constant activity. Much in the Chinese countryside depends upon the course of events year by year in this important field.

Commodity Grain Bases: Surplus, Self-Sufficiency and Deficit of Grain

Farmers in most parts of rural China are expected to contribute to local contractual obligations for supplies of grain to the state. In addition some areas are designated as 'commodity grain bases', which are expected to be regular suppliers of grain in quantity, though Chinese lists and maps of these vary a good deal (Chen Dunyi and others 1983, 90; Geographical Research Institute 1980, 121; Sun Jingzhi 1988, 234). Figure 5.5 represents a broad consensus on the location of the grain bases based on these authorities.

One reason for indecisiveness on the identity of the grain bases is no double their very different characters. The most reliable are of three kinds. One kind (like Jiangnan, at the mouth of the Yangzi), has both high population density and very high outputs; a second kind (in the north-east) has medium-to-low yields but low population densities; the third kind (in the middle Yangzi provinces) has medium-to-high yields but medium population densities. The less reliable also come in three different kinds. In Sichuan and the Pearl River delta in Guangdong the recent growth of population and

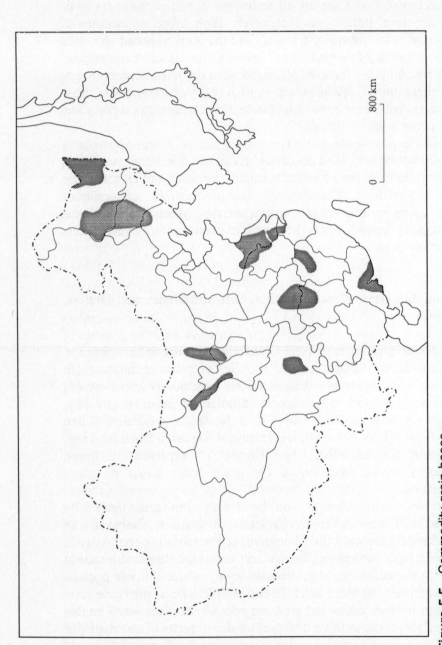

Figure 5.5 Commodity grain bases

The locations of commodity grain bases are agreed among Chinese writers in general terms, but different detailed versions are published. Those shown in this map are a composite version.

Source: Chen Dunyi and others 1983, 90; Geographical Research Institute 1980, 121; Sun Jingzhi 1988, 234.

800 km

commercial demand for grain have practically consumed the former surpluses. On the great bend of the Yellow River, inland in Ningxia and Inner Mongolia, a grain base exists which is designated more in terms of its relative strength (in a deficit area) than its absolute capacity; and the same applies to its extension towards Central Asia in Gansu. Finally, there have been a number of references to an intended grain base, with a variety of actual locations, in northern Jiangsu and eastern Anhui, where there has been extensive land reconstruction since liberation, based on rationalization of the river system.

Of these grain bases, it is clearly those which are most valuable, because they are close to major population clusters, that are the most vulnerable to depletion and final disappearance of surpluses owing to population growth and growing economic activity. In these terms, those of the north-east and the middle Yangzi seem likely to have the longest lives. But it may be argued that the concept of 'grain bases' itself is a throwback to Maoist thinking, and dependent on the idea of grain squeezed at low prices out of passive countrysides. When the state finally allows the price of grain to rise freely, regional coercion can hopefully be abandoned.

Grain bases are important, however. In 1988 the state purchased some 50 million tonnes of grain under contract, about 12.5 per cent of the total harvest. Of this, around 52 per cent was contributed by 100 counties each supplying more than 160,000 tons, and nearly 40 per cent by only 58 counties which each supplied more than 200,000 tons (Deng Yiming 1991, 21). These 158 counties (7.2 per cent of all counties) are the backbone of the whole procurement system. They have not been individually named, but most are certainly within the grain bases.

When all the relevant factors are taken into account, China's grain surplus and deficit provinces are those shown in figure 5.6. The self-sufficient regions are those of the far west, Ningxia and Shaanxi in the centre–north, and Hebei and Shandong on the north China plain. Deficit regions are the rest of the north, the three great cities, and the advanced provinces of the east and south coasts. Surplus regions are the north-east and the Yangzi provinces plus Henan at the southern end of the north China plain – this is the main region of supply to deficit areas, partly because of its advantages in shipping grain by water transport. The total deficit is

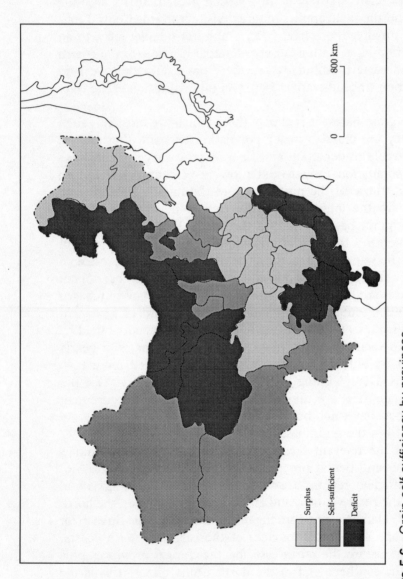

Figure 5.6 Grain self-sufficiency by provinces
Deficit areas include some which have excellent farm conditions but very heavy demand, and some where natural conditions are poor. Most surplus provinces have good conditions and only moderate population densities. Self-sufficiency in the north China plain was first proclaimed under the Maoists in 1974; it represented a considerable achievement then, and still does so.

Source: Constructed from material in Academy of Agricultural Science 1990

not large – around 3 per cent of total output. Of this, about one-third is due to shortages in poor and pastoral regions like Inner Mongolia, and most of the rest arises in cities and mining or industrial regions such as Shanxi. However, figure 5.6 does not tell the whole story. Important transfers arise within the groups of provinces, for instance of rice from Hunan against wheat from Henan in the north China plain, or of soya beans (reckoned as a grain) and maize from the north-east against rice and wheat from provinces to the south. An oddity in the data from which the map was constructed is the difference in treatment of Sichuan (as surplus) and the south-east coast provinces (as deficit). In fact both regions have lost their former status as regularly in surplus in recent years, due to such factors as loss of land to local industry and use of grain to feed pigs or make wine; both are now irregular performers close to self-sufficiency. In the text, however, Sichuan's commercial conversion of grain is explained sympathetically, but the south-east coast is advised to mend its ways.

The Agricultural System since 1979 – Achievements and Problems

The golden age

Agricultural reform was the spearhead of the whole reform movement after 1979 – partly because of the widespread disenchantment with official management in the countryside but probably more directly because of the countryside's capacity for self-management, and hence capacity to fit into the communist party's emerging concept of a non-command, but non-capitalist, economy. Self-management was the central paradigm. 'Contract' and 'responsibility' systems of landholding were introduced in 1979, and by 1981 some 90 per cent of villages throughout China had adopted them (Leeming 1985a, 52).

Before the reforms the brigade's land was managed as a kind of collective estate, where local workers were in effect employees. Since the reforms, the land remains in the ownership of the collective (now usually the village), but it is broken up into parcels, and individual families are able to 'contract' for the management of particular parcels of land, to provide for their own substance but under the obligation of particular deliveries (typically of grain) to the collective. Not only arable land, but also other village resources, such as ponds, plantations, poultry and animals, are open to contract by local people. This system is the heart of rural resource management in China up to the present. Among a capable, responsible peasantry who understand their own land and their own production systems, it has self-evident merits: the assurance of motivation; stability and flexibility in the face of natural disasters like flood and drought; and the transfer of practical farm decisions to individuals who depend upon them.

The reform system at first maintained the state's monopoly of trade in grain, but official procurement prices were substantially increased. In addition some of the capacity of the officials to interfere with rural management was dropped – in part explicitly, in part by implication.

Impressive increases in production were obtained, particularly in grain outputs. Previously, in spite of heavy increases in fertilizer and pesticide use in the 1970s, grain output remained stubbornly slow to increase – by only 3 per cent per annum (at constant prices) between 1971 and 1978. In 1980–5 grain output increased (also at constant prices) by 7 per cent per annum (*SYC* 1986, 130). This seems to be partly the outcome of improved motivation due to the new system, partly the delayed result of the increased technical outputs that became available in the 1970s, but which had little effect because of official interference with agriculture requiring grain to be planted everywhere, often in unsuitable locations. In Mao's time, for all the effort put into grain outputs, no annual grain output exceeded 300 million tons; but in the first half of the 1980s the reforms succeeded in making grain plentiful, at least for a time, and the harvest of 1984 was the first to exceed 400 million tons. Total sown area fell, and so did the tractor-ploughed area, but fertilizer input continued to increase, and so, impressively, did grain yields. For grain as a whole, output was 1635 kg per hectare in 1965 and 2335 kg in 1978, but by 1984 the figure was 3615 kg, a percentage increase in 1978–84 of 43 per cent. For wheat, the percentage increase in yield in the same six years was 61 per cent. At the same time, peasants were encouraged to diversify and specialize in their production systems; official hostility to private marketing in local towns or at tourist resorts was given up, and peasants were able to plan their day-to-day activities, as well as being responsible for their own production systems and livelihoods. Peasants became in effect small tenant farmers.

For some years, under these conditions, rural China seemed to have entered a golden age, of busy and successful farmers enjoying security, progress and rising standards of life. The urban free markets, closed in Mao Zedong's time and now re-opened, provided once more for urban people the immense variety of rural produce which Chinese cooks need. Partly by contrast with the grey years under Mao, the early 1980s experienced a remarkable sense of revitalization in the countryside.

Towards a new crisis

Golden ages, as we know, rarely last long without coming to decay. By 1984 the rural system introduced in 1979–80 had come to full fruition. In that year grain output reached 407 million tons, generating uneasiness in the Chinese media about storage and the proper use of such a vast harvest; output of cotton in 1984 also reached more than 6 million tons. The harvest of 1984 was not significantly exceeded until the 420 million tons of 1990; grain production in recent years has generally hovered around 390 million tons, cotton around 4 million. Where 407 million tons was thought to be an embarrassingly generous harvest, it might be thought that a somewhat lower figure would be welcome, but that is not so. Demand for grain as animal and poultry feed, for bakeries producing cakes and pastries, and for making beer and spirits, remains high and growing; and population itself has risen by close on 100 million people between 1984 and the present. The rural golden age ushered in by the reintroduction of family farming has faded out. Commentators in the media are now once more struggling with the problems of rural China. We are reminded that throughout human experience, the rural producer who thinks food prices are high enough has not been invented, nor the urban consumer who thinks they are low enough.

What are these problems – what became of the glowing promise of the early 1980s in the Chinese countryside? The heart of the matter is the sluggishness of grain output, and it will be best to start with that problem. There are several strands to the argument.

In 1985 the state abandoned the obligation of local units to grow 'quota' grain for official procurement, and instead introduced a system of contracts between the official purchasing organizations and individual peasant households, normally through the local village units. At the same time the state reconstructed its price system for contract grain. Grain was now to be purchased by the state at premium prices for the first 70 per cent of contracted quantities, and at negotiated prices for the rest and for production above contract. This reversed the previous arrangement, and appears to have reduced farmers' incomes from grain sales by about 10 per cent (Hou Zhemin and Zhang Qiguang 1989, 38). It appears also to have led farmers (who are now in effect managing their own farm businesses) to plant crops other than grain. However, officials in many rural areas treat the new grain contracts as effectively equivalent to the old quotas, and still represent the state's demands as obligations (Oi 1986), although this is administrative action without legal foundation. The system has a marked tendency, no doubt predictable, to operate each year on the basis of last year's precedent; at least at first, some peasants had not even heard that the system had changed. At the same time, the state has also given up its monopoly of trade in grain and the obligation to purchase all the grain produced by farmers. However, the problems of grain sales to the state do not end with the payment system. Payment itself has been a problem. In successive years since 1985 the state agencies have adopted the practice of issuing IOUs to peasants in payment for grain, rather than cash. Needless to say, this practice is extremely unpopular. State agencies have sometimes gone to the length of requisitioning grain when peasants were unwilling to part with it, and in other cases speculators have been able to move in, paying cash for grain when the state could not (Li Shusheng 1989). In yet other cases, the state has resorted to various kinds of administrative blackmail to force peasants to deliver grain – refusing to allow children to go to school or enter college, or seizing livestock, cash or savings bonds instead (Liu Zifu and Wang Man 1989). On the other hand, if grain is plentiful, the state may default on its contracts. Around 1984 and 1985, there was much optimistic talk of the capacity of grain specialist households (who were expected to have larger-than-average landholdings) to provide for most of the local contract obligations for grain deliveries, but these have dropped out of prominence, partly no doubt because of difficulties with payments; perhaps also because such households might equally work in more profitable alternatives, such as cash crops or rural industry. Official grain prices have been raised again in 1992, perhaps as part of a move intended to bring them up finally to market levels, and in Guangdong grain prices were freed.

Problems with grain farming are not restricted to those involving the state. It has been widely argued in China in recent years that costs of growing grain have increased more rapidly than grain prices. Costs of inputs such as fertilizer have risen much faster and further than grain prices. One typical response by grain farmers has been to use less fertilizer. Meanwhile for farm products other than grain (cash crops such as sugar, oilseed rape, animals and so forth), prices have risen much more rapidly. It may be argued that since the communist party has now adopted a rural model which gives prominence to specialist and diversified farming, it is not unreasonable to expect outputs other than grain to be favoured, but of course grain supplies are increasingly needed, not only as staple food but for bakeries and other specialist uses, for making beer and spirits, and also for animal and poultry feed. In these conditions, there is an increasingly sharp difference between the market price of grain and the price paid by the state (Sun Zhonghua 1990). As a result of lack of peasant enthusiasm for grain production (other than for their own household supplies), the phase of rapid increase of yields per unit area between 1978 and 1984 has been followed by a phase in which yields have fluctuated around the figure for 1984.

There are other weaknesses in the grain system. Fertilizer and diesel are considered expensive; in addition, they may be hard to obtain. Investment on the land is limited, as shown in chapter 4; and most peasants in the present phase would certainly claim that farm costs are now so high that serious capital investment is out of the question. Local officials may be much more preoccupied with township enterprises, which are profitable and prestigious, than with the problems of farmers.

The problems of high costs in grain farming, and consequent low motivation among farmers, have been partially tackled in various ways, notably by the continued use of profits from rural industry to subsidize grain farming – 'industry subsidizing agriculture' as the communist party likes to call it, but more frankly 'rural subsidizing urban'. In places with strong rural industry, this relationship was already established in the 1970s (Leeming 1985a, 118). It is bound to relate constructively to the present regime's policy of increasing commercialization of rural outputs, but it is no less bound to guarantee proletarianization of grain agriculture and the peasants who practise it.

Peasants in China are now supposed to be free to put their labour into other kinds of cultivation, but of course many peasants, especially those far inland and distant from the cities, lack experience and skill in farming other than grain; in addition, as will be shown, specialisms such as pig farming generate their own problems, sometimes dependent on grain.

Specialisms in the Countryside

Grain scarcity apart, can the salvation of the countryside be found in specialisms of various kinds, tapping the new prosperous urban markets for items such as poultry?

Much less has been written in China on this topic than might have been expected, and the general tone is not optimistic. Aubert (1990, 24) shows that outputs in this field (pork, eggs, milk) have continued to increase since 1984, and he explains this by the free market rationale. In fact prices of these items rose each year at rates of 5 per cent or higher – in 1988 by 37 per cent.

Growth of total output per annum between 1984 and 1989 has been 9 per cent in pork, 13 per cent in eggs and 15 per cent in milk. What is disconcerting, however, is the observation that outputs per peasant household have hardly moved at all in this time; for eggs, the figure has actually fallen (*CSY* 1990, 369; *SYC* 1986, 165). This is surprising in a situation where demand is strong and prices have been rising quite rapidly. Study of the provincial breakdowns of these figures (*CSY* 1990, 372ff.; *SYC* 1986, 168ff.) shows that most provinces have shared the same experience, though outputs per household throughout tend to be much higher in the south, and there has also been stronger growth in the south. In these conditions it is perhaps not surprising that the middle-class urban people who write in the media are dissatisfied. What is perhaps surprising is the scale of the apparent sluggishness of incomes from varied production, in conditions where demand remains strong.

Pigs

The most important of the specialist items is pork, and in fact pork is notorious in recent years for a succession of supply-and-demand cycles (Aubert 1990, 25). Pork is the universal Chinese meat, and often the subject of official intervention. The state monopoly of pig purchase was given up in 1985, but local official units such as provinces and cities still interfere in the market with subsidies for consumers or producers, usually according to short-term rationales. It is argued that when the price relationship of grain to live pigs by weight is around 1:5, pig production develops steadily and pork prices are stable. But since 1979 this relationship has fluctuated a great deal, partly because of official interference, partly because of farmers' intuitions about the market, guided by the changing price of grain and the costs of fodder for pigs. Since 1985, the cost of rearing a pig to marketable size has risen to a point where it may exceed the pig's value at slaughter. A variety of subsidies, which differ from province to province, can affect both the price of pork and the cost of rearing pigs – and meanwhile, organs of government at all levels may be heavily in arrears in subsidy payments (Wang Ping and Song Qing 1990). The practice of putting pork supplies into cold store does not appear to help much. Some writers argue for more, and much more purposeful, official intervention, pointing out that Chinese pig-rearing is extravagant in feed use and other ways by international standards; others start from the reality of the 180 million peasant households that each raise a pig or a few pigs, and argue for freedom from official intervention which often generates or compounds instability.

Cotton

Another specialism of particular importance is cotton, which exceptionally is still the subject of a state monopoly – cotton cannot legally be sold except to state organs. Cotton output was held at a low level under the Maoists, but after 1979 was encouraged to rise rapidly by official price increases – though in recent years, because of rapid increase in the use of artificial fibres, cotton is no longer as cardinal in Chinese clothing as formerly. The most important indicator and determinant for progress in cotton is comparable with that for pigs: the level of price per kilogram of cotton as against grain (Song Qing 1990). Arrangements in the early 1980s gave a ratio between grain and cotton prices of about 1:10.5, against figures such as 1:9 in the 1970s. It is argued that stability and healthy growth in cotton output require a ratio around 1:10.2 in prices between cotton and grain. In 1984 cotton output was very high; but as with grain in the same boom year, the state over-reacted and introduced much less favourable purchase terms for cotton in successive years, falling to 1:8.2 in 1988. At this kind of price (less than cultivation costs, it was said) cotton became scarce. Grain procurement prices rose by 76 per cent between 1984 and 1989, but cotton by only 36 per cent, and cotton output fell in 1989 to 39 per cent less than in 1984.

The cotton shortfall has inevitably had industrial consequences (Li Tiezheng 1990). Everywhere major state cotton mills have experienced the greatest difficulty in obtaining supplies. Some mills purchased illegal cotton at high prices outside the state's monopoly. Needless to say, in these conditions the demands of township mills were criticized particularly bitterly; these mills compete with state mills for supplies, but of course they are consistent with government policy for rural diversification, and of course have their own dependents. Some peasant households are said to be hoarding cotton.

Under the Maoists, the state plan had to authorize the planting of cotton, and cotton cultivation (which was more profitable than grain) was made conditional upon the locality's being self-sufficient in grain. One result of this was the reduction of cotton production in the north China plain (highly suited to cotton, but with dense populations and short of grain), and its increase on the Yangzi in such provinces as Jiangsu and Hubei, often using various intensification systems such as intercropping (Leeming 1985a, 167–70). In the 1980s, the authorities have loosened this policy, and allocation of land to cotton (still an official responsibility) has been made much more to north China plain counties than previously. The results are shown in general terms in table 5.1, where Shandong,

Table 5.1   Production of cotton – main provinces, 1981 and 1989
(thousand tons)

	1981	1989	Percentage increase
China	2968	3788	28
Shandong	675	1025	52
Hebei	222	536	141
Henan	355	527	48
Jiangsu	563	485	−14
Hubei	353	313	−11
Xinjiang	114	295	159

Sources: 1981 from SYC 1981, 150; 1989 from CSY 1990, 352

Hebei and Henan all register substantial increases in output, but
Jiangsu and Hubei show reductions. The return of cotton to the
north China plain restores its importance as before 1949. Interestingly, the largest increase of all is registered by the central Asia
province of Xinjiang, with excellent cotton conditions.

Vegetables

Vegetables are one of the most important of all groups of crops in
China – each person is reckoned notionally to consume about a
pound ($\frac{1}{2}$ kg) of vegetables daily; and there can be no doubt that the
great variety of vegetables grown in China helps out greatly with the
creation of variety in Chinese diets. The list includes practically all
kinds grown in Europe and North America, plus others which are
local and characteristic – Chinese celery, Chinese 'chives' and
various kinds of cabbage and gourd. Most vegetables are grown, as
in traditional times, close to their market, on the outskirts of towns
and cities.

It is not surprising that the official vegetable supply system in
Mao's time, handling perishable commodities in very large quantities, and with supply governed in part by the weather, did not
perform well. The production systems were loosened in 1981,
substituting contracts for quotas, and the state monopoly of
vegetable marketing was given up in 1985. Peasants now produce

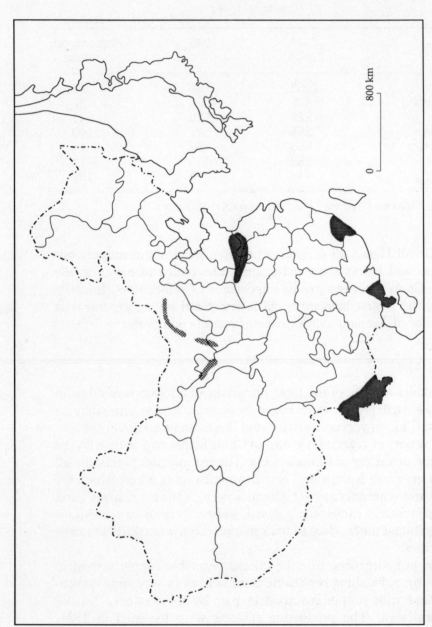

Figure 5.7 Vegetable bases
The map shows the actual and potential 'vegetable bases' mentioned in the text.
Sources: see text

vegetables under contract at official prices, and also outside the contract system at market prices which are roughly double (Shi Wei 1990). Families growing vegetables under contract receive production subsidies and also grain and coal at state prices, and their risks in production and marketing are virtually nil, being assumed in effect by the official marketing system. Vegetables are becoming scarce again within the official system, however, because of farmers' family members leaving the land and giving place to hired labour, because of a tendency of farmers to use less fertilizer than formerly, because some farmers are giving up vegetable growing altogether, and because farmers increasingly sell their best and earliest vegetables on the open market, keeping only inferior crops for the contract.

Meanwhile there has been discussion in China about the development of vegetable 'production bases'. It is argued (by analogy with California or Spain) that vegetable production could be concentrated increasingly in places which have a climatic advantages. Most vegetables prefer conditions which are cooler and dryer than south and central China in summer, and most do not survive the cold of northern China in winter. Already there are important regions of commercial vegetable production, most or all of it not contracted, which supply distant cities. Such areas include south Fujian, areas in the sub-tropical southern and western valleys of Yunnan, and the southern Leizhou peninsula of Guangdong (figure 5.7). These areas obviously produce what are really luxury supplies, destined for well-off people in the major cities, and for all the areas named there must be formidable transport burdens involved, though quantities are not yet large – 100,000 or 200,000 tons per year from each area. In addition an important area within which vegetable outputs for the northern cities have been extended through several counties is on the Henan–Shandong–Jiangsu border (figure 5.7). This area now ships out a million tons of vegetables annually (*Jingji Cankao* 17 April 1989). Some writers (Zhong Xiaopo 1989) have suggested that in addition to supply bases in the south, a number of areas on the edge of the high plateaus might be developed to improve summer supplies (figure 5.7) – areas in Gansu and on the 'inland delta' of the great bend on the Yellow River in Ningxia and Inner Mongolia. From a production point of view this is surely feasible; but here too transport (around 1500 km from Lanzhou to Beijing) is

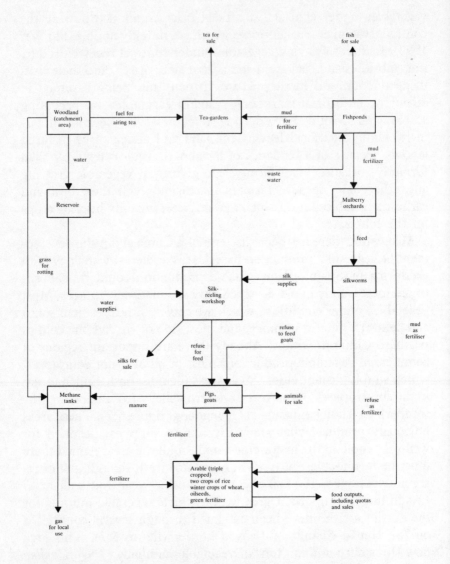

tea for
sale

fish
for sale

Woodland (catchment) area) → fuel for airing tea → Tea-gardens ← mud for fertiliser ← Fishponds

water

mud as fertilizer

waste water

Reservoir

Mulberry orchards

feed

grass for rotting

Silk-reeling workshop ← silk supplies ← silkworms

water supplies

silks for sale

refuse for feed

refuse to feed goats

mud as fertiliser

Methane tanks ← manure ← Pigs, goats → animals for sale

refuse as fertilizer

fertilizer

feed

fertilizer

Arable (triple cropped) two crops of rice winter crops of wheat, oilseeds, green fertilizer → food outputs, including quotas and sales

gas for local use

Figure 5.8   A complex production system from Zhejiang
This is one of a series of similar systems published around the same time in 1983–4. They represent ideals of intensification in the old style of the first half of the century, but improved in efficiency and increased in scale. Little has been said about such systems since official policy took the route of commercialization and shedding of labour in 1985.
Source: Ma Xiangyong and others 1984, 58

a forbidding problem. However, carriage of vegetables over long distances is already common within the state marketing system; a recent article about supplies at Changsha in Hunan mentions suppliers as far afield as Inner Mongolia and Hangzhou – though regrettably both in the context of rotten consignments (Zhang Du and Liu Pingchun 1988).

Complex production systems

One ideal in the field of specialism is the complex production systems of the kind illustrated in figure 5.8. In this scheme the arable land, woodland, tea-gardens and mulberry orchards are the foundation of a complex series of food chains involving fish, pigs and goats, silk and tea outputs. This scheme comes from an advanced village on the edge of the mountains in Zhejiang, some 40 km inland from Hangzhou (Ma Xiangyong and others 1984). Its complexity is interesting, though probably in places idealized. But what is still more interesting is the fact that this scheme already looks seriously dated. The three arable crops were always very hard to manage successfully; they were a Maoist ideal rather than a practical proposition. A village of this kind would now be certain to have at least one or two factories, making perhaps bamboo furniture or even knitwear. It would now be unlikely to be as self-sufficient as the Maoist ideal suggests; it will certainly need more woodland inputs than it is probably able to supply. Tea-gardens, mulberry orchards and animal herds will now be the subject of contracts with families, and so of course will the arable land (Leeming and Powell 1990, 163–7).

Rural Industry and Other Rural Enterprises ('Township Enterprises')

According to official figures, only about 45 per cent of the 'rural total output value of society' in China comes from agriculture. Forty-one per cent is contributed by rural industry, and the remaining 10 per cent by rural construction, transportation and trade (*CSY* 1990, 316). Rural industry must necessarily be taken very seriously according to these figures. Even if the importance of industry is exaggerated by the use of figures for output values (as a reflection of low prices for grain), this exaggeration itself (operating through the effects on cultivation of high-paying as against low-paying work) is of importance in the community. In fact, of the rural labour-force, 79 per cent is occupied in agriculture, 8 per cent in industry and 13 per cent in other kinds of work (*CSY* 1990, 313–14). Obviously a degree of instability is suggested by the disparity between these figures for inputs of labour and outputs in terms of value. Needless to say, there are marked differences in the importance of rural industry among provinces. In Liaoning, Jiangsu, Zhejiang, Hebei, Shanxi and the three great cities, rural industry is worth more than agriculture, although all these except Shanxi have above-average agricultural outputs per rural worker. Meanwhile in backward provinces such as Yunnan, Guangxi, Inner Mongolia and even Xinjiang, outputs of rural industry are less than 20 per cent of those of agriculture – in some cases much less. There are now around 15.4 million rural enterprises of various kinds, employing nearly 94 million people – both of these figures are for 1989, and indicate some decrease since 1988, due to financial stringency. The most important kinds of business are shown in table 5.2, which distinguishes between importance in terms of output and of employment. Since 1985, rural businesses both collective and individual have been grouped together as 'township enterprises'. About one-half of total employment in enterprises is contributed by individual (*getihu*) businesses, but only about one-third of total output. Most rural factories are collective, owned and managed by rural towns or villages. Individual businesses can

Table 5.2   Rural businesses in four categories, 1989 (percentages of total)

	Output values	Employment
Manufacturing	74	54
Building	12	25
Transportation	6	10
Commerce/catering	8	11

Source: *CSY* 1990, 316, 313. These figures appear to include official business.

be of any kind, but are more usually in trade or transportation; many are local shops.

Output per employee is higher in the latter two categories than in manufacturing, in part at least reflecting their relative scarcity. The preponderance of manufacturing among these enterprises appears to reflect two main considerations: the preponderance of town and village enterprises, which are usually factories; and the continuing unwillingness of individuals to start businesses such as transport or commerce, or to expand them to a point where they might attract adverse notice among local officials. In fact the preponderance of manufacturing has tended to increase since 1985, when a new deal was proposed for township enterprises, and local officials were given incentives to encourage them. Between 1979 and 1985 commerce and building (but not transport) grew more rapidly than manufacturing.

For manufacturing in collective ownership, figures are given in table 5.3. These headings cover around 70 per cent of such industry under either heading. It is not surprising that industries with a high skill content (machinery) have higher outputs per worker than those with bulk outputs (building materials). The average enterprise employs about 30 people, but family-based enterprises employ less – about 13. In 1985, rural enterprises of all kinds employed a total of 64 million people; in 1989, more than 90 million. But industrial distribution does not appear to have changed very much.

Table 5.3 Rural industry under collective ownership, 1988
(percentages of total)

	Output values	Employment
Building materials	16	29
Textiles	13	9
Machinery	11	8
Metals	11	9
Food and drink	8	7
Chemicals	5	3
Energy	3	5

Source: Chen Yaobang and others 1989, 104–5

The Maoists claimed rural industry as part of the agricultural economy and enlisted it in their struggle for more grain. Since 1979 it has become increasingly evident that the strongest rural industries are often those which are least 'rural' in character – particularly those which are clients of urban factories. The harsh reality behind this situation is that urban factories are often short of space, water, cheap labour and official permission to expand, especially for dirty or space-consuming processes. Rural locations are much less restrictive. Moreover, the differences between urban and rural incomes are still sufficient to make low urban wages look highly desirable to farmers. In addition, industrial work continues to enjoy more prestige than farming, partly because it is thought to be cleaner; and factory work often requires a shorter working day. It is true that in these circumstances urban pollution is often being exported cheaply to the countryside and the rural economy proletarianized, but neither the officials nor the rural people see much harm in this. 'Free-standing' local industries, based on local resources and management, are usually free of these faults, but they are often less stable than client factories – lacking by comparison capital, technology, markets, official contracts and official protection.

There are dramatic differences among provinces in the importance of local enterprises of all sorts. The strong provinces in this respect are of course those with coastal locations, relatively skilled workforces and relatively highly developed managerial talent (see figure 6.3). Aside from the three great cities, which all have very high figures, table 5.4 suggests degrees of difference which arise between advanced and backward provinces.

Table 5.4   Gross values of local enterprises per person employed, 1989 (yuan per person)

Advanced provinces		Backward provinces	
Guangdong	13,589	Henan	6599
Shandong	11,165	Hunan	7024
Jiangsu	12,943	Gansu	4684
Liaoning	12,254	Yunnan	5905

Source: CSY 1990, 382, 383

Chinese writers now comment freely on the experience of rural industry, and much of the comment is adverse. Rural industry is criticized for its customary low productivity, for the habit of tax evasion and hence (it is said) unfair competition with state industry, for wasteful use of energy, water and raw materials, for pollution which may be serious, and even for its contribution (through local subsidy to agricultural costs) to the continuing cheap urban food system in China. Most local industry is owned by local units such as villages and towns, which displeases the out-and-out protagonists of private enterprise; and the same people argue (not without reason) that services such as transport businesses are what the countryside needs, rather than factories. Collectivists meanwhile argue that privately owned local services and businesses are also often exploitative, and even more prone to tax evasion. Collectivists tend to be sympathetic to village industry, but also defensive of state industry when faced by rural and small-town competition. Free market enthusiasts are suspicious of the official side of township enterprises and their contribution to the continuing survival of the collective system.

The Status of Rural Industry and Diversified Production

It is tempting to argue that the rural production systems have been – so to speak – hijacked by rural industry. Table 5.5 presents some relevant figures. Increases in the contribution to rural outputs by industry have been by almost ten times in the nine years 1980–9, while the increase in the value of agriculture (including animals and sidelines) has been only two-and-a-half times. Notoriously, industry is the great generator of local prosperity in rural areas. But within agriculture itself something similar has happened to crop farming by comparison with non-farming activities, as table 5.5 shows. Crop farming itself has increased outputs by only 166 per cent in these same nine years, while branches other than crop farming have made more than 400 per cent growth. Values of individual crops are not given, but it is clear from all the qualitative evidence that, in the rural production systems, grain prices are the least mobile.

Table 5.5 Contributions to rural outputs, 1980 and 1989
(thousand million yuan)

	1980	1989	Percentage growth 1980–9
Rural total output values of society	279	1448	419
Gross outputs of agriculture	192	653	240
Gross outputs of crop farming	138	367	166
Gross outputs of branches of agriculture other than crop farming (forestry, animals, sidelines, fisheries)	54	286	430
Gross outputs of rural industry	54	589	991
Gross outputs of other rural activities (construction, commerce, transport	33	207	527

Source: CSY 1990, 316, 318

The provincial distribution of these changes in rural outputs other than farming can add something to an understanding of them. Provinces with the highest increases in rural industry are primarily those of the north China plain, plus a few peripheral cases such as Gansu and Guizhou. High-performance provinces such as Jiangsu, Zhejiang and Guangdong have generally lower figures for growth, partly because in 1981 they started from higher bases. On the other hand, provincial figures for animals, sidelines and so forth since 1981 give average figures for north China plain provinces, but above-average for advanced provinces in the south such as Guangdong, Zhejiang and Jiangsu. These southern provinces have of course better environmental conditions for all the various outputs involved, such as much better water supplies, warm winters with more wild vegetation for animal feed, and more abundant and more varied woodland. But something does seem to depend on market access: the figure for Sichuan (with similar physical advantages, but less urban demand) is below average.

The general picture that emerges from these figures is one which is becoming familiar. Backward areas are usually those far inland. The north China plain, by far the most important backward area although by no means the most backward, is now taking an increasingly full part in Chinese rural development, primarily through industry. Advanced rural areas in the south have enjoyed continued development, though no longer at rates which are markedly stronger than those of the north China plain. Agricultural opportunity other than for grain, however, remains more varied and positive in the south than the north. This finding agrees with detail (*CSY* 1990, 370–1) which shows that outputs per rural household of animal and sideline products are generally higher in the south than the north – the average number of pigs per rural household in eight southern provinces is 2.9, against 1.1 in the north China plain.

Limitations in Peasant Society

In some degree, the peasant enthusiasm for migration to town in China, in the face of official discouragement and many specific hardships, must be an expression of disappointment at the achievements and capacity of rural systems to create the kind of life people seek. Surveys of migrants have indeed indicated as much. Both economic and social weaknesses of rural systems are quoted, but particularly economic. It is true that in the present phase rural families can reconstruct their livelihoods in various ways, but rather few do so, either as farmers or as entrepreneurs. Why is this so? Little has been written on this point, but there has been some study and thought.

On factor is said to be the traditional outlook of honouring farmers, especially grain farmers, and despising merchants and others who make money through skill and expertise. 'All people here are genuine peasants' was one reported remark. Most rural people think in terms of household self-sufficiency, it is argued, and so do many rural cadres. Ignorance is widespread; 36 per cent of rural people are said to be illiterate or semi-literate (Gao Ping

1990) – and of course in most Chinese villages there is not much to read. Maoist outlooks, echoing the traditional negative view of business, still inhibit potential entrepreneurs and are still adopted by officials. 'Some cadres cannot tell the difference between those who become rich through exploitation and those who become rich through labour.' Bureaucrats frequently put obstacles in the way of local people who wish to sell produce, or refuse to deal with outsiders or allow others to do so. Some officials exploit well-off peasant households by insisting on meals at their homes and by extortion (Beijing broadcast 1984, K13).

Specifically, peasants find that they are the victims of a long list of fees and charges levied by officials at various levels – for the birth control programme, for the militia, for schools, for the social security system, for administration costs, for local investment. Charges of this kind, levied at will by officials at various levels, together with tax, cost the average peasant between 10 and 15 per cent of his notional net income – but these charges are levied in cash, while only about 65 per cent of the peasant's income comes in cash – sometimes much less.

A New Generation of Rural Problems

Apart from the weaknesses attendant upon grain agriculture, a new group of problems is emerging in the Chinese countryside, especially structural problems for which there is no evident solution. There are essentially problems of limited supply and unlimited demand in the resource field – or, to use old-fashioned language, problems of man–land relationships.

Land itself is becoming scarce in eastern China. There is still a vast amount – millions of hectares – of under-utilized mountain and hillside land in these regions, but accessible lowland, which is normally of good quality, is already scarce, because of demand from state and city occupancy (usually involving requisition and change of ownership), peasant house-building, erosion losses, losses due to brickworks, claypits, gravel pits and open-cast mining, local industry and so forth. Burials are now again on the increase. Self-evidently there can be little economic development without use of land. Most of this land was previously arable; some of it is of the highest quality, by world standards as well as Chinese. Cultivated land per person fell, according to official figures, from 1733 square metres in 1957 to 1066 square metres in 1977 (Leeming 1985a, 66). This figure has fallen again since 1979; the rate of loss of farmland during the 1980s has been of the order of 670,000 hectares (6700 sq. km) annually, according to a high official in the land administration (Wang Xiaokun 1989). Farmland per person by the year 2000, says a critic, will amount to only 973 square metres – equivalent to a square plot with sides some 30 metres long (Xiao Jiabao 1989).

There are also problems in the field of land allocation and investment on the land. One result of family farming under contracts has been a return to extreme fragmentation of farmland among peasant families. Moreover, it has already been pointed out that investment in Chinese agriculture has usually been quite limited, even under the Maoists when the state professed particular rural sympathies. The misgivings of left-wing critics of the present phase, that family farming would weaken investment on the land still further, seem to have been justified: investment on the land is now at a very low ebb. When official investment does take place in present conditions, it often takes the form of major development projects intended primarily to increase outputs of commodity grain, cotton, edible oil and so forth (*China Daily* 28 November 1989, 1). Investment by individual farmers in their own production systems, with or without official support, remains scarce. Production continues to be manged on a year-by-year basis and generally as economically as possible. This is partly because production systems are perceived as having little or no surplus; partly because farm families still have little confidence in the permanence of their present status; partly because farmers do not feel that they can afford extravagances. One result of farm costs rising faster than farm prices is a tendency of farmers to cut costs whenever possible; reduced inputs of fertilizers are one of the reasons for falling yields, especially of grain.

Rural population meanwhile continues to grow, with its necessary demands for food, jobs, houses and incomes of all kinds. Continuing rural population growth, as well as urban, is one reason for the state's continuing anxiety about grain outputs. In the present phase the 'babies' of the baby boom of 1962–72 are young adults. There is already a problem of surplus farm labour in the countryside, usually estimated as around one-third of all rural labour, which has been compounded by improved labour motivation due to the restoration of family farming.

In 1984–5 the communist party moved towards a long-term solution of some of these problems, with limited measures whose effects have been powerful. The central authorities' new plan for the countryside proposes a sharp break with both traditional and Maoist forms of crowded labour supplies and continuing intensification. The new scheme proposes basically commercial countrysides with fewer farmers and larger landholdings, more specialization, less labour and more capital, more machinery, more conversion of grain into other kinds of food such as poultry and eggs, much less self-sufficiency, much more business (Chinese Communist Party 1984, 1985) Reconstruction of the grain procurement system belongs to the same group of changes. In this context, two quite limited relaxations turned out to have momentous consequences – toleration of migration of rural labour to jobs in town, and increased encouragement for rural industry. Rural industry has been discussed in an earlier section, and rural-to-urban migration is discussed in chapter 7. But there is room here for some points which relate to the effects of migration on the countryside.

Farmers who leave their contracted land do not usually surrender their contracts, but the land may well be uncultivated, or it may be sparsely cultivated by a brother or father who has no labour to spare. Many people who take jobs in local industry cannot farm their contract land properly. However, surrender of contract land for reallocation is rare because the contracted land is perceived as a kind of security – and in a country where official policy has been so volatile during the past 40 years, individuals feel the need to retain security of any kind. It should also be recalled that migration deprives the countryside of labour. Where so much labour is said to be surplus, this may appear unimportant; but it is less than clear that this labour is surplus within the present cultivation systems, which are essentially labour-intensive. Migrants are usually men, and often men of an age to be running their own farms. As always, it is much easier and quicker to abandon old obligations than to construct new ones. Where redistribution of farmland has taken place in the villages, we hear, this has usually been to accommodate the new generation of adults, and has typically resulted in falls in output (Sun Zhonghua 1990). Production of specialist outputs has grown rapidly, but not steadily. There is still widespread fear of a reversal of policy at the top – indeed the principal effect of the Tiananmen Square disaster of 1989 in China as a whole has been to remind the community of the capacity of the authorities for ruthless action. From the authorities' point of view, this is good political thinking; but inevitably it inhibits people from taking up economic adventures, as well as political.

SIX

The Industrial System and the State Economy

Industrial Growth

The industrial system is one of the great success stories of modern China under the communist party, and perhaps the most important of all. It is generally recognized that modern industry was tiny in scale and backward in technology in 1949, and that in most respects China's industrial system is now commensurate with her size and resource endowment. These are immense achievements. But as in most success stories the achievements have been gained at costs of various kinds. Above all, it is now argued that the basic assumption of the industrialization policy – that what was good for Chinese industry was good for China – was flawed in various ways.

As already shown in chapter 4, the People's Government has treated the state industrial system with particular favour throughout its rule – or, to put it in another way, the powerful industrial bureaucracy has normally had a great deal of its own way. Industrial investment as a percentage of all investment was 44.6 per cent in the vital First Five-Year Plan period, and it has rarely fallen below that figure in subsequent phases (table 4.3).

It is true that before 1949 China had little industry. 'Manchukuo' or Manchuria, the present three provinces of the north-east, was a Japanese protectorate between 1933 and 1945, and in that condition received substantial investment. Some of this was stripped by the Soviet authorities in 1945, but most of China's heavy industry in 1949 was in Liaoning province, and inherited from the Japanese phase. Shanghai was the great centre of light industry, especially textiles. It is also true that at that time most Chinese industry occupied coastal locations, though this was not only because of imperialist contacts and foreign priorities. Even before the Opium War of 1840–2 and the opening of the treaty ports with their institutionalized foreign contacts, China's most advanced economies and most of her industry were in coastal provinces, as has been shown in chapter 3.

Figures are available to illustrate the scale of industrial growth since 1952 (*1949–84 Statistical Materials*, 51–2) – a growth of nearly 1000 per cent in coal output; of nearly 3000 per cent in steel; of nearly 5000 per cent in electricity; of 35,000 per cent in artificial fertilizer, and so forth. Provincial figures for outputs before 1981 are very few, but if the simple distinction between 'coastal' and 'inland' is adopted, in many branches of industry (including all those mentioned) the great bulk of expansion took place inland. Achievements on this scale in forty years are hard to exaggerate.

The development of industry in China since 1949 has been much more continuous than that of agriculture, and much more closely tied to Soviet precedent. Although the course of industrial development under the First Five-Year Plan (FFYP) of 1953–7 was considered far from smooth at the time (Hughes and Luard 1961, 48–55), the momentum and broad strategy of that plan were maintained in essentials into the 1980s. The FFYP strategy proposed the priority development of heavy industry, especially iron, steel, coal and heavy engineering, as the heart of the industrial economy, together with the opening of a series of new industrial centres in the interior of China. Some two-thirds of existing industry was already nationalized in 1953; the Plan assumed the development of industry for the future as a central function of the state. The Plan document itself (which was published only in 1955) is quite secretive about details, and gives only very limited indication of the rationales of these features of strategy, but they appear to have been adapted both as matters of principle and as practical steps towards the creation of a more rational and realistic economic geography; one which made better use of important mineral deposits, which could support the revitalisation of the deeply depressed northern half of China, and which would be better protected from attack from the Pacific.

Understandably within this scheme, and in spite of the Soviet removal of equipment there at the end of the Second World War, the Japanese installations in the north-east (primarily in Liaoning province) were taken as one major foundation. Those of Shanghai, however, were not; Shanghai was relegated to the peripheries of the plan. Alongside developments in Liaoning at Anshan, Fushun, Shenyang and Luda, the FFYP also proposed expansion of heavy industry at a number of other centres, all inland and mostly in the northern half of China. New 'key steel bases' were to be located at Baotou in Inner Mongolia, Taiyuan in Shanxi and Wuhan on the Yangzi in Hubei – all supplied by local or relatively local mineral deposits, and all intended to function as generators of industry of other kinds. Other centres nominated for development in the plan were Xian in Shaanxi, Lanzhou in Gansu, Loyang in Henan and Chongqing (Chungking, the wartime capital of China) in Sichuan.

It is hard to quarrel with the geographical side of this broad strategy. All the cities named were old-fashioned official and business cities which badly needed industrial functions. All were located in stagnating agrarian (or part-pastoral, in the cases of Baotou and Lanzhou) countrysides which badly needed technical and economical revitalization, fresh sources of employment and stimulus towards new cycles of development. All have profited greatly from official investment in the years since 1953; all are now major regional industrial centres. Xian, for instance, depressed, backward and almost without modern industry in 1949, has a vast complex of electrical industries, started under the FFYP, which continues to develop up to the present and now employs upwards of 20,000 people. For cities of this kind, the Plan has represented the critical introduction of whole regions to the technical possibilities of the twentieth century. The rationalization of the time was that by these means 'consumer cities' were reconstructed as 'producer cities'. A close (though not always happy) symbiosis developed between the communist party and the cities, which continues to the present in such forms as subsidies for urban food supplies. Significant rural-to-urban migration took place, both to supply factory workers and to work on the very large infrastructure projects created by new industry, new housing estates and the demand for higher standards, especially of hygiene, as will be shown in chapter 7.

Most of this industry was heavy industry – an imprecise term but one whose meaning came increasingly to be 'not for sale to small-scale or individual consumers'. Industrial producers produced increasingly for industrial consumers, suggesting a new meaning for 'consumer cities'. The main purchasers of state industrial outputs, usually at officially manipulated prices, were other state units of various kinds. This inward-looking relationship fostered continuous growth of industry by offering reliable markets and suppliers, but it resulted in a broader consumer market starved of supplies and an industrial system deeply dependent on its system of feather-bedding. In addition, the development of the Chinese industrial networks between 1958 and 1979 took place under remarkably secretive conditions imposed by the state, with virtually no attempt to rationalize decisions publicly in either structural or locational terms.

There has been a price to pay. Subsequent planned industrial developments have typically followed the logic of the first Plan, with emphasis on heavy industry and on a number of 'key' industrial units in the regions. Industry, especially heavy industry, became the life-work of a powerful industrial bureaucracy, self-serving and self-perpetuating. Production itself remained self-serving, capable of providing other parts of the privileged heavy industrial network with parts and materials but unable to supply needs further down the line. Supplies of steel for agricultural machinery, building and consumer use were never adequate, and remain inadequate to the present. Bureaucratic practice, joined with the effort of survival in the Maoist upheavals, made industry prosperous but left its development hopelessly tied to official precedent, especially FFYP precedent.

Apart from Shanghai, Wuhan, Chongqing and Guangzhou, there is still little large-scale or heavy industry in southern China; most of the big projects are in the north. In addition Chinese industry has remained closely tied to the great cities – some old cities, like Xian, were greatly stimulated by the new industry, and shot further into prominence as a result; but others, such as Tianjin, were already prominent. Industry has remained essentially urban, much of it tied to the great cities. In the 1980s 42 per cent of state industrial output came from 17 major cities with aggregate population only 4.5 per cent of China's total (Leeming 1985b, 419). Industrial output is now more widely distributed, but not much more widely. The remarkable but still obscure Third Front policy of around 1964–76, which funnelled industrial and related investment into a group of centre-west provinces centred on Sichuan, seems from this perspective to have been a revival of FFYP thinking in an extreme form (Cannon 1990, 36–9). The new steel city of Panzhihua, in Yunnan close to the Vietnam border, seems to be the most tangible result of this policy, but the steel output of Yunnan remains only around 1 per cent of national output.

Strengths and Weaknesses of the Official Systems

China's state factories (those 'owned by the whole people') number about 100,000 (*CSY* 1990, 390). In addition to these, there are also 'collective' factories, belonging to major administrative units such as cities and counties. The latter are technically not 'state' factories, and they are supposed to be guided, rather than controlled, by the state plan, but in practice the differences between big collective factories and state factories are not great, particularly in respect of the problems of reform in the present phase. In China as a whole there are only some 12,300 'large and medium-sized' factories (ibid.) and virtually all these are in either state or collective ownership and management. In 1989 the state sector produced 56 per cent of gross industrial output and the collective sector 5 per cent (*CSY* 1990, 393, 394); the rest came from township and individual enterprises. Taken together, the two major official sectors still represent 61 per cent of Chinese industry, and they form the backbone of both the urban and industrial systems. Through payment of profits and taxes, they also form the backbone of urban and state revenues, though subsidies to loss-making state enterprises are a heavy drain on the same revenues.

It is consequently of the highest importance, especially to the communist party and the high bureaucrats, how the industrial economy performs. Recent Western opinion is that it performs only very indifferently – Tidrick and Chen (1987, 1) argue that 'the fundamental weaknesses of China's economy have been inefficient use of labour, capital, energy and raw materials and a mismatch between the composition and quality of production and demand'.

This seems fairly comprehensive. The essence of the industrial reforms since 1978 suggests that Chinese official opinion sees the industrial systems in much the same light: witness policies for the reversal of the customary priority for heavy as against light industry; the recognition of consumer needs; the reduction of extravagant use of energy; the increase of exports; the upgrading of existing plants and methods, rather than new capital construction; the creation of incentives to efficiency; the decentralization of control; and the substitution of market relationship for administrative controls. All these are reforms of kinds associated with market systems rather than socialist systems. It is not surprising that some of them have run up against built-in resistance from the old political and administrative system which is maintained by the communist party: the commitment of the industrial bureaucracy to heavy industry; their ignorance of conditions in foreign markets; the continuing effects of managed prices within the industrial system, such as low prices for coal; the continued effects of state or other collective ownership of factories, and so forth.

Incentives have been one problem. Before the reforms, state factories handed over their profits to the state organ responsible for the factory. This deprived the central treasury of revenue and operated as a disincentive to efficiency in the factory. In 1983 this system was converted to separate out a higher tax take (to go to the state treasury) and a degree of profit retention by the factory (Leeming 1985b, 423). However, not all retained profits were sensibly used; in the hotel boom in the middle 1980s some factories used retained profits to build hotels. In 1987–8 contract systems were introduced, based on the concept of those used in agriculture, but necessarily much more complex. Understandably the variety of these systems is very great; understandably also, the right terms for such contracts have been difficult to devise. In practice, using contracts fixed on the basis of past successful operation, the most efficient factories may find themselves driven the hardest by unrealistically high targets set by the bureaucrats. Contracted allowances for costs may be rapidly overtaken by price rises. Official policy changes, like the reduction of investment since 1988, may greatly modify an enterprise's operating conditions without any possibility of compensation; and the same is true of all kinds of administrative interference (Fang Dahao and others 1990). Perhaps for these reasons the state in 1990 nominated 234 'key' state enterprises to be run in the old way, clearly within the state plan, with the state as guarantor of supplies and purchases, and as collector of profits as well as taxes (*Jingji Cankao* 20 May 1990, translated in JPRS *China Report* 26 July 1990, 20–1).

Another typical and deep-rooted problem is organization. Before the reforms, economic units such as state factories followed administrative and communist party precedent in accepting regulation of junior units by senior. Connections outside these 'vertical' strands of authority (for instance, between parallel units in neighbouring counties) were unorthodox and rare – and pointless, because activity (procurement, marketing) was supposed to take place only through the proper channels. Efforts since the reforms to promote fruitful lateral connections have had only limited success, because administrative structures remain powerful and because many people who manage them are reluctant to relax control (Hodder 1990). And where much continues to depend upon mutual back-scratching between bureaucrats, lateral relationships which do develop may well be no less uncompetitive than the old vertical connections.

Local protectionism is really one aspect of the same problem. At the provincial level, protectionism relates to the self-interest of some provinces in trying to protect provincial production of, say, bicycles, by taking steps to discourage the 'import' of better bicycles from Shanghai or Tianjin. The state is strongly opposed to this kind of local policy, partly because it runs counter to present economic thinking, partly because it might feed local separatism. But even at the local level, it might well be in the interests of a county to protect the local market for its own output of fertilizers; many fertilizer factories at county level have been running at a loss. Many problems of this kind arise within the industrial systems at all sorts of levels – procurement of dyes or patterns, machinery and machine parts, steel of various qualities, and so forth. 'National' markets are exceptional. Although trade does take place between enterprises in widely separated provinces, markets may also be difficult to enter, where supplies are scarce and each potential supplying enterprise already has its regular list of customers.

Investment has its own problems. Attempts are being made to substitute official loans (for which interest is paid) for state grants for industrial capital. Only the state and its organs are rich enough to finance big-scale investment. The state bureaucracy's perceptions may well be limited and old-fashioned; but factory managers' and local bureaucrats' perceptions may be biased towards short-term profit. Experiments in shareholding have taken place in China, but so far these have been of the debenture kind, which guarantee a fixed income for a period plus the return of capital. There is no equity stock or trading in shares, and realistically little or no risk.

Prices in industry remain a minefield. Outputs produced 'within' the state plan are subject to official allocation at official prices, and ought to be manufactured using official materials. In principle raw material prices are low for Chinese enterprises but prices of manufactured outputs are high; understandably, as a long-term result investment has been slack in the raw material industries, including energy, while manufacturing investment has been abundant. But since 1979 the pressure of pent-up consumer demand has been too great for industry to satisfy; hence consumer prices have risen steeply, tearing industry still further from its Stalinist roots. In present conditions not all the output even of state factories falls within the state plan, or is at controlled prices. But official prices are still structured to favour business within the state plan and official objectives generally. Unofficial activities hence usually operate at higher prices and costs – the 'double-track' pricing system. The system works because of 'official' scarcities and rigidities which continue to stimulate the unofficial side of business.

The industrial planning system creates its own problems. Policy in industry has stressed quantitative growth to the neglect of specific market demand, even demand within the industrial system; as a result quality is often low or simply inappropriate to need. Similarly, product mixes have tended to follow official prejudice rather than market need. Meanwhile, low-grade or non-specific outputs are being over-produced. In part this is a natural problem for an industrializing country in a generation like the present, with extremely varied technical needs; but the planning system has not helped.

Many of the problems that have been outlined are among those which give the township enterprises their opportunities. Township enterprises do not depend upon the state bureaucracy, and the local officials upon whom they do depend have their own needs for profitable and growing industries. Nothing (except local official obstruction) need hold township enterprises back in seeking commercial contacts of any kind; and these enterprises are often quick to seize black-market opportunities in the state sector – for instance, for supplies of diesel. For local industries, the rigidities in the state system (including those of price and supply) can create their best opportunities; and local officials may well be zealous in exploiting them, to the disadvantage of official units higher up the administrative scale.

Industrial Regions

Figure 6.1a shows provinces where industrial output per inhabitant is greater than the national average. The area shown includes the three great cities, all the north-east, and the coastal provinces except Fujian, Hainan and Guangxi in the far south. No other inland provinces are indicated. The companion map (figure 6.1b) shows provinces with less than two-thirds of the average of industrial output per person – nearly one-half of all provinces, mostly with values around 55 per cent. All (except Guangxi and Hainan) are inland provinces. It is interesting that there are very few provinces with figures close to the average – those nearest are Hebei (adjacent to Beijing and Tianjin) and Hubei (including the great industrial city of Wuhan).

Figure 6.1 suggests a familiar generalization which of course runs counter to most policy since 1949 – that industrial development is strongest in the coastal provinces. However, the provincial framework is a crude one, and a more exact approximation is possible. This is proposed in figure 6.2. This suggests two main industrial areas in China, one in the north including Beijing and Tianjin, most of Hebei and Liaoning, and parts of Jilin and Henan. It produces about 23 per cent of Chinese industrial output, but from a very extended area nearly 1200 km in length. The other lies in Shanghai, southern Jiangsu and northern Zhejiang. This area produces around 15 per cent of Chinese industrial output, of course from a much smaller area than the first. A third major kind of industrial area is

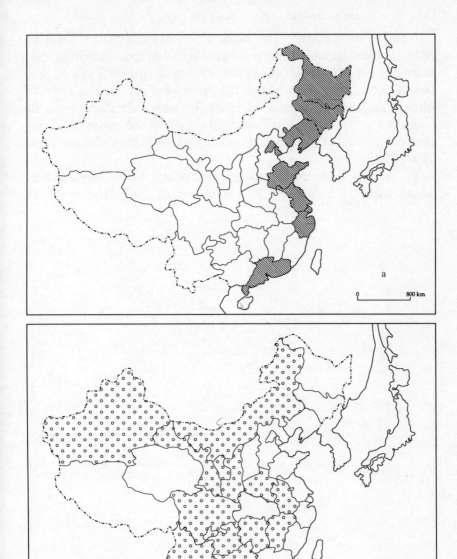

Figure 6.1 Industrial output per person, 1989
a, Provinces with industrial output per person greater than the
national average; b, provinces which score less than two-thirds of the
national average.
Source: *CSY* 1990, 396, 83

also shown on the map. This is the cities throughout the country. Many of these lie within the 'industrial areas' already identified, but many do not. Those which do not and which appear in figure 6.2 have joint industrial output of the order of 9 per cent of China's total. The total of around 60 cities for which data are given in Chinese sources have joint industrial output of the order of 40 per cent of the total (Urban Social and Economic Survey Organisation 1990, 228–9).

To identify industrial areas has a corollary: to identify areas in which industry is scarce. This is above all the inland south of

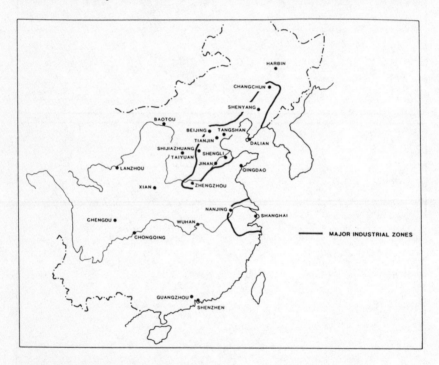

Figure 6.2   Chinese industrial zones
The two large areas shown contribute around 38 per cent of industrial output, and the cities shown which lie outside these two regions contribute an additional proportion of around 9 per cent. Important areas which lie outside those shown and which do not lie in cities include coal outputs in Shanxi and oil in Heilongjiang, and important local industries in such provinces as Guangdong, Shandong and northern Jiangsu.
Source: Leeming 1985b, 419

China. Eight southern provinces (Hunan, Jiangxi, Sichuan, Guizhou, Yunnan, Guangxi, Fujian and Hainan; not including Hubei or Guangdong) together produce only 16 per cent of industrial output, against 31 per cent of China's population. This is the most important 'underdeveloped' region in China, apparently little modified by investment said to have been made in the 1960s and 1970s under the 'Third Front' programme. The region's status no doubt represents an opportunity for future generations, but obviously industrial weakness for the present.

Problems of Productivity in Industry

In most respects, the scheme of industrial development proposed in the FFYP appears to have remained official orthodoxy from 1953 to 1979 in spite of radical changes in social and economic policy in 1958, 1962–3 and 1966. Industry remained inward-looking, preoccupied with captive industrial markets. Industrial prices were set high enough to enable even relatively inefficient producers to remain in business (Lardy 1978, 158). The official industrial decentralization movement of 1957 actually strenghtened centralization, as the *People's Daily* said at the time (Lardy 1978, 148).

Since the start of the reform movement in 1979, there has been much more official interest in the coastal areas, with their inherent comparative advantages (better-educated work-forces, more experienced managers, proximity to other producers and to markets, advantages of international contacts), but official support for industrial growth inland has continued up to the present. There seem to be several reasons for this continued interest in the interior, of which administrative inertia is the least justified but not necessarily the least important: recognition of growing congestion in eastern China, recognition of the interior's primacy in the supply of industrial energy, recognition of the sub-continental scale of China and her development problems, especially transportation problems. However, strategic considerations, which were influential in both the FFYP and 'Third Front' phases, do not now appear to be prominent.

The difficulty with industry in the interior is low productivity. About 39 per cent of Chinese industry reckoned by value of output is located in inland provinces; but reckoned by industrial employment, this figures rises to 46 per cent (*CSY* 1990, 392, 396). And low productivity per unit of input is not only a problem of labour, as table 6.1 shows – it affects capital and energy as well. In the table only eight of the thirty provinces consistently out-perform the national average in productivity – the three great cities, together with Shandong in the north, Jiangsu and Zhejiang in the centre, and Fujian and Guangdong in the south. All except Beijing have coastal locations. A few inland provinces, such as Hubei with its important industrial metropolis of Wuhan, produce creditable figures for one or two indicators, but realistically there are no other provinces which come close to those named. What is most surprising in the table is the low standing of the north-east, especially Liaoning with its long industrial tradition; but in China old-established industry tends to be state-owned, and so to have suffered over the years from force-feeding with both labour and capital, and accompanying starvation of innovation. Liaoning occupied much the same place in the figures for outputs per worker (the only ones available at that time) in 1982 (Leeming 1985b). Dramatic change since 1982 is most evident in the cases of Guangdong and Fujian, which have shot into prominence in the new phase of enterprise culture, partly with the help of capital contributed by and through the overseas Chinese communities.

To what extent have the changes of the 1980s modified the geography of Chinese industry? This question will be tackled in terms of some specific outputs in the next section; here an attempt will be made to respond in broad terms. Table 6.2 gives some basic figures.

The answer is: not greatly, but with some consistent movement. Of the three official macro-regions (East, Centre and West), the East seems to have gained slightly over the Centre and West – this would usually be attributed to the increasing advantage of skill and expertise in the present system plus more effective growth of smaller industries and much more support from abroad. The north China plain has gained slightly in strength – more than slightly if the loss of priority by Beijing and Tianjin is taken into account. Shandong has been particularly successful. All three of the great cities have lost priority, due no doubt to problems of congestion. Finally, the eight strongest provinces are defined in both years at a

Table 6.1   Productivity in Chinese industry, 1989

	Outputs per worker (thousand yuan)	Outputs per unit of capital (yuan)	Outputs per unit of energy (yuan)
China: averages	16.6	140	2891
East region (coastal provinces)			
Beijing	25.2	155	2732
Tianjin	23.1	159	3236
Hebei	13.9	123	1271
Shandong	19.3	145	2702
Shanghai	28.3	199	4608
Jiangsu	20.0	224	4184
Zhejiang	18.8	241	4926
Liaoning	15.3	117	1647
Guangdong	23.7	187	3984
Guangxi	14.8	140	1992
Fuijian	17.2	179	3350
Hainan	14.3	98	n.a.
Centre Region			
Hubei	17.0	134	2278
Hunan	13.3	139	1538
Jiangxi	11.8	139	1942
Jilin	12.6	112	1513
Heilongjiang	12.2	97	1531
Henan	13.2	116	1541
Shanxi	11.3	86	876
Anhui	13.1	150	1795
Inner Mongolia	9.9	82	978
West region			
Shaanxi	13.7	106	2000
Gansu	14.1	84	1111
Ningxia	12.6	80	1325
Sichuan	12.6	124	1748
Guizhou	12.6	96	1190
Yunnan	15.7	119	1553
Tibet	7.0	39	n.a.
Xinjiang	13.6	80	1147
Qinghai	12.4	60	1091

Note: All outputs are gross outputs. Outputs per worker, 1989, from *CSY* 1990, 423. Outputs per unit of capital, 1989, ibid., 426; figures are outputs per 100 *yuan* at original values. Outputs per unit of energy, 1989, ibid., 479 (converted); national average from ibid., 393, 464; figures are outputs per ton of standard coal. Figures do not appear to include township or individual enterprises. The three regions named are those shown in Figure 4.4.

Table 6.2   Industrial output values by individual regions, 1981 and 1989 (percentages)

	1981	1989
East region–coastal provinces	61	63
Centre region	27	25
West region	13	12
North China plain (Beijing, Tianjin, Hebei, Shandong, Henan)	23	24
Beijing, Tianjin, Shanghai	20	13
Eight strongest provinces	56	58

Sources: SYC 1981, 214; CSY 1990, 396

point representing 4.6 per cent of total output; in 1981 these eight provinces contributed 56 per cent of total industrial output in aggregate; in 1989, 58 per cent.

Industrial Production in the Regions

The spectrum of Chinese manufacturing is now quite wide, and of course in terms of location the range of factors involved is quite varied from one product group to another. Some of these will now be explored (CSY 1990, 442–9).

Investment

Some outputs show clear signs of being based upon large-scale state investment over several decades, and hence of being basically direct outcomes of successive state plans. Motor vehicles furnish the clearest example – obviously a motor works requires heavy investment and a reasonable total scale (table 6.3). The motor industries in Beijing, Changchun in Jilin and Wuhan in Hubei all date from

Table 6.3   Motor vehicles, 1981 and 1989

	1981		1989		Percentage growth in totals
	Thousands	Per cent	Thousands	Per cent	
China	176		583		231
Beijing	27	15	85	15	215
Jilin	60	34	84	14	40
Hubei	40	23	127	22	217
Liaoning	4	3	49	8	1125
Others	44	25	116	41	164

Sources: SYC 1981, 245, 250; CSY 1990, 437

the FFYP phase, and those in Beijing and Wuhan have maintained their position during the vast growth of the 1980s; Changchun has not. The many (upwards of 100) motor vehicle assembly plants which were set up locally in the long phase of scarcity of vehicles in the 1970s and early 1980s produced on a relatively small scale, and often at low levels of efficiency. 'Others' in the table also include (by 1989) joint-venture factories such as Volkswagen in Shanghai (now said to rank with Changchun and Wuhan), Peugeot in Guangzhou, Daihatsu in Tianjin and the American Motor Corporation's venture in Beijing (Ko and Liu 1990). All these foreign initiatives have taken place under the aegis of the state. None is yet large enough to compete with the big state firms; Tianjin, the biggest of the newcomers, has only 7 per cent of output. However, Liaoning has strengthened its position and Jilin has lost ground, and the contribution of provinces other than the 'big four' has risen from 25 to 41 per cent. Most of Chinese motor vehicle production continues to be in buses and lorries rather than cars, and partly for this reason there was substantial import trade in motor cars from Japan, until the crackdown on spending in 1989–90.

Coastal locations

The most important aspect of the regional change in industry is that suggested already by the figures for productivity – the differential growth of industry in the coastal provinces. Much of what is said in

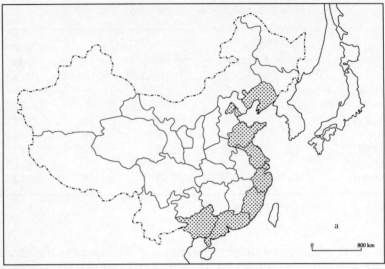

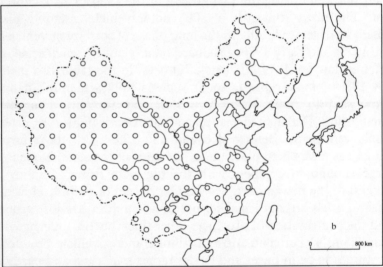

Figure 6.3 Township enterprises – outputs per rural worker, 1989
a, Provinces with outputs per rural worker higher than the national
average. Except for Hebei province, this is a complete 'coastal'
distribution; the figure for Hebei is quite high but not marginal. b,
Provinces with less than half the national figure, including almost all
non-coastal provinces – only Jilin and Heilongjiang, Shanxi with the
coal industry, and Hubei have modest but below-average outputs.

Source: *CSY* 1990, 383, 313 (the figures do not include outputs by individual (*getihu*)
businesses)

the sections which follow relates to this broad development, for instance in the case of steel, but it can best be illustrated from high-tech industries because of their need for a skilled and well-educated work-force and managerial class – or from the low-tech point of view of the township enterprises, because these are started in places where enterprise is most readily available. Both of these kinds of condition are satisfied best in the coastal provinces.

In high-tech industries, China has 'Asian' advantages such as those enjoyed by Hong Kong. Fifty-seven per cent of Chinese camera production (which has assumed significant proportions only since 1974) comes from the three great cities plus Jiangsu and Zhejiang. For sewing machines, production is even less widely distributed. Shanghai, Guangdong, Jiangsu and Zhejiang together produce 65 per cent of output (*CSY* 1990, 443, 445). When consumer durables like sewing machines were still very scarce in the early 1980s, many other provinces opened factories to manufacture them (Liaoning, Hunan, Henan), but in many cases outputs have since fallen, as competition bit later in the decade. Cameras and sewing machines are both significant export items.

Figure 6.3 shows the distribution of township industries in terms of total output per rural worker. The high-performance and the low-performance groups are equally basic documents of progressive social distributions in China – all coastal provinces except Hebei (and the new Hainan) score high, but Hebei scores close to the average. Other relatively high scorers are Hubei (always an advanced province by the standards of the rural south) and Shanxi with its village coal-mines. Very low figures are those for the central Asian provinces, the centre-north (Gansu, Inner Mongolia) and the south-west (Sichuan, Guizhou).

Dispersal of industry and growth of new centres

At the opposite end of the spectrum are a number of commodities whose production is mainly official and planned, but for which investments were made very widely during the 1960s and 1970s, partly in acknowledgement of transport and supply difficulties, partly in acceptance of a degree of devolution of economic decision-making to the provinces. This has been the case with steel. As may be envisaged, this kind of development revised, or at least comple-

mented, the FFYP insistence on raw material-based heavy industries located inland or in the north-east (table 6.4). Since the 1960s,
Shanghai itself has had a major steel complex at Baoshan – though
it should be added that further development at Baoshan is now
somewhat controversial. Liaoning, prominent in steel since the
1940s, is still the principal producer, followed by Shanghai; but
both have lost precedence recently as others' steel outputs have
grown. FFYP producers like Sichuan, Hubei and Inner Mongolia,
who contribute similar outputs to the increased total, have also lost
precedence, and now belong to the third rank. Their place has been
taken by provinces which belong to what might be called the newly
industrializing north – that is, the north China plain, in this case
notably Shandong and Henan. This kind of productive growth is
taking place within the state systems and mainly or entirely within
the state plan, and it represents, as it would in the West, differential
developments in investment, costs, congestion, management (or
official) initiatives, and so forth. But change in these terms cannot
be rapid.

Table 6.4   Changes in steel outputs, 1981–1989

	1981		1989	
	Million tons	*Per cent*	*Million tons*	*Per cent*
China	36		61	
Rising status				
Beijing	2	6	4	7
Shandong	1	3	4	7
Henan	1	2	3	5
Hebei	2	6	4	7
Falling status				
Liaoning	9	25	11	18
Shanghai	5	14	5	8
Sichuan	3	9	3	5
Hubei	3	9	3	5
Inner Mongolia	3	5	4	4

Sources: CSY 1990, 435; SYC 1981, 241

Domestic production, however, is not the whole story with steel. China now ranks fourth among world steel producers, but steel is an important import – apparently owing to China's limited output of steels of special quality.

Output of cloth seems to follow a similar rationale. In addition to provinces which have long histories in the textile business (Jiangsu, Zhejiang, Shanghai), a number of provinces now contribute important outputs, often based on long-standing state investments. These are particularly Hebei, Shandong, Henan and Hubei, all provinces with important cotton outputs. Details are not given, but it would appear likely that garment production would display a similar distribution and rationale.

Local demand and supply

Some kinds of production approximate closely to regional and provincial populations – cement is a good example. Only the south-west region produces much less cement than its population would suggest, and the north China plain, north-east and Lower Yangzi slightly more than the population percentages.

Glass appears to follow a similar rationale, and so do some chemical outputs. Plastics also display a similar distribution, and the same is true of some other outputs, such as pesticides. It must be borne in mind that provincial output in a particular field may well depend upon a single factory.

Organization within the State Industrial System

It is sometimes imagined in the West that the management of an industrial system whose main component factories all belong to the state, or its organs such as cities or provinces, should be relatively simple – and that allocation of supplies by official agencies under the state plan should also be a relatively simple matter. Nothing could be further from the truth. 'The state' is a body with many arms. Not all 'state' factories are controlled by the same arm; some are directly within reach of more than one. Changes over time have often left complexes of arrangements. In addition, practical industrial organization involves many different bodies and forms of relationship.

In the field of supply and marketing, Tang Zongkun (1987, 210–36) outlines a scheme which is rich in complexities and ambiguities. Commodities, including industrial raw materials, are handled at three administrative levels according to the official view of their importance; a factory must usually make several kinds of requisitions – to provincial, city and ministerial and local agencies, for example. Prices within the plan are fixed and low. However, not all these requisitions are necessarily met, which leads to bottlenecks in production and habitual exaggeration of need in requisitons. The requisition system takes account primarily of quantities, often with little reference to quality of inputs provided, whether in respect of timber for building, or more technical inputs such as dyes for textile works or steel for engineering works. The allocation system works slowly; enterprises must place requisitions on the basis of estimates of need within a detailed plan which they have not yet received. Deliveries within the planned system are often erratic, either because suppliers fail to carry out their own planned production or because their suppliers do not produce necessary components – or because of difficulties on the railway system. Transport difficulties are exacerbated by the planned linking of production centres which are distant from one another – one example given by Tang (1987, 220) is an engineering product whose manufacture involves two intermediate steps in transportation, each of around 1500 km, equivalent to Manchester–Marseilles–Hamburg. In these circumstances, enterprises sometimes hoard supplies, and a variety of deals can be struck, to the benefit of enterprises but the weakening of the system, for instance by barter. This is one way in which the growth of the local industrial sector (which lies outside the state plan) may help state enterprises, although barter deals are irregular and, strictly speaking, illegal, and rail transport outside the plan may be difficult to arrange. Moreover, prices outside the plan are usually higher than official prices, sometimes much higher.

It is the business of the enterprise to supply its outputs to the state organs, which will allocate them to consumers, also at fixed prices that are low. When outputs are in surplus, free marketing of the surplus is allowed, though presumably official knowledge of a surplus in one year may affect allocation levels in the years following. For outputs such as steel and motor vehicles, however, the scale of demand outside the plan relates to macroeconomic conditions, both generally and in the official parts of the economy in particular – for instance in construction.

How do these various special features of the industrial system relate to China's geography? To some extent the relationships are adventitious, as whether the planned progression of production involves distantly separated production units, or units in the same province or region. In these cases, development depends upon the perceptions of bureaucrats who function within a closed system. If there is a consistent rationale in this respect, it is probably one based on an old-fashioned kind of economic geography, in which perceived resource or entrenched investment generate specific kinds of production. Such a rationale is bound to be based on precedent year by year, to say nothing of the bureaucratic advantages of following precedent; it seems bound to be unresponsive to changes outside its own system. To this must be added the system's tendency to growth, where growth is also guided primarily by precedent. In practice, the missing 'mobile' part of Chinese industry seems to be growing up in the jungle of the township enterprises, in spite of the powerful official component in the ownership and management even of these.

It is often argued that as long as major industry continues to be owned by the state or by state organs, official interference with production and exchange is inevitable. This is probably true at some level, but not necessarily at the present level of official interference. The heart of official management of industry at present seems to be less state ownership than official interference through the mandatory plan. Part of the need for official interference derives from the inheritance of Stalinist systems of manipulation of prices, always to the advantage of state management; but the whole problem of prices in China is one which the communist party still lacks the will to tackle.

SEVEN

Urban China

The Scale of Urbanization

China in 1949 had a low level of urbanization, reckoned at 10.6 per cent. The urban population was then around 58 million out of a total of 540 million (Kirkby 1985, 107). The historic empire also had a low urbanization level. A few cities, led by Shanghai, had more than 2 million people in 1949, but most ordinary provincial capitals had closer to half a million (Hangzhou, Harbin, Changsha); and many of these had grown startingly during the Japanese War (Sit 1985, 6, 28–9). The Chinese communist party began its revolutionary career in the 1920s among the industrial workers of Guangzhou (Canton) and Shanghai; it was the armed force of the bourgeois republic that drove the communists to seek survival among the peasants. Partly by reason of the form of that survival, partly for other reasons which arise from Marxist thinking about the cities, the communist party has had a long history of ruralist rhetoric. But as Kirby (1985, chapter 1) points out, the reality of communist practice has been otherwise, mainly because of the party's insistence upon industrialization and the continuing growth of industrial production at least as far down the hierarchy as the county capitals. It is the industrialization imperative that has shaped China's urbanization, not abstract notions such as anti-urbanism (Kirkby 1985, 14). Even in the phase 1961–76, when policy insisted upon the removal of some 20 million people from the cities (many of them in-migrants of the late 1950s, attracted by industrial expansion), the continuing demands of industrial expan-

sion resulted in recruitment of peasants from neighbouring countrysides (Kirby 1985, 119). Since the mid-1980s there have been
major developments in the cities, including widespread building
and a good deal of migration whose relations with state industry are
not direct, but this is new, and still not really orthodox.

Kirkby (chapter 3) outlines the problems of definition and
unexplained changes in definition, which plague study of Chinese
urbanization through population totals. The 1989 official figure for
urban population is 573 million, 51.5 per cent of the total population (*CSY* 1990, 89, 90), but this figure relates to all persons living
within designated towns and cities. Criteria for designation of
towns were changed in 1984; the 'urban' population figure for 1983
was only 241 million, 23.5 per cent (*CSY* 1988, 75). However, since
1984 important changes have also taken place on the ground,
notably the dramatic rural-to-urban migrations which are outlined
in chapter 7. The figure for urban population derived from the
census of 1990, of 297 million, is surely realistic (*People's Daily* 31
October 1990, 3). What is particularly important is the recognition
that in the present phase China is experiencing rapid and large-scale
urbanization akin to that which has been experienced in most Third
World communities during the past generation, and for much the
same reasons. Migration produces large supplies of cheap labour,
which is important for the township enterprises and the informal
and service economy generally. It may also suggest new dimensions
of social instability, especially in the great cities.

The Urban Landscapes

Traditional urban landscapes in China were densely packed,
crowded with people and business, and rich in street uses such as
hawking, cooking and eating. In all these respects they differed
little from those of other parts of Asia, or indeed those of Europe
until the nineteenth century. Under the communist party the cities,
particularly the city centres, changed more than any other part of
the visible landscape of China. Communist party hostility to private
business led to the effective nationalization of first wholesale, then
retail, trade, and the ideal of collectivization led to the closing of
individual workshops in all sorts of trades in favour of officially

organized collective industrial units. Even hawkers were cleared away, partly in the name of tidiness and hygiene in the streets (as in contemporary Singapore), partly in the name of hostility to the principle of private enterprise, partly through the official allocation of supplies which left none for small-scale individual users. As a result of these changes, which gathered momentum during the 1950s and on into the Cultural Revolution movement of 1966 and subsequent years, Chinese city centres lost up to 90 per cent of their restaurants and other businesses, most or all of their entertainments, virtually all their hawkers and all their markets except those run by the city (Leeming 1985a, 23–4). Wholesale and large-scale business was now conducted in offices in compounds rather than in shops on the street, and increasingly such business was done by telephone. Small-scale business was usually managed by the former owners as employees or partners of the city authorities, but understandably such a system had little vitality, and with the passage of time these businesses tended to die. In consequence, in the Maoist years, city centres were quite dreary, with scanty and predictable Chinese-made merchandise, very little choice and (particularly in the official markets) openly contemptuous service to the public. Fruit and vegetables of good quality seldom appeared: those that existed went to restaurants and canteens which enjoyed official privilege. One justification for the ubiquitous bicycle in Chinese city centres has been the lack of anything for pedestrians to look at or to do in the streets, except go to work.

Some of this has changed in recent years, particularly since the middle 1980s. Private businesses, and traditional-style signs advertising them, have reappeared in the streets. Hawker stalls are again tolerated, even encouraged. There is much more merchandise in the shops, some of it imported, much of it oriented towards middle-class customers. Private cars are still practically unknown, but there is much more traffic, particularly business traffic, in the streets. Local clustering of small businesses like cobblers on street corners has returned. Night-markets, which are reconstructed each evening and sell mostly clothing, occupying either daytime food-market sites or ordinary business streets, have reappeared. Official food supplies have lost their automatic superiority; in some cities the main restaurants, managed by the city authorities, still have stale rice, slow service and predictably fewer customers than less pretentious privately run establishments in street stalls; in other places the city may have refurbished and reorganized its restaurants to take their share of the new consumer prosperity.

The present return to a more traditional scheme of city-centre (and suburban) marketing and servicing began after 1979 with a system of 'free markets'. At first, free markets occupied streets designated by the city authorities for marketing by peasants and other local producers of items (mostly food) needed in the city; these were normally side-streets close to the city or town centre, but not the main thoroughfares. The free-market streets are usually still very important in the new central business systems, but the original scheme has widened both geographically and in terms of commodities until much of the city and town centre has been reoccupied by commerce. Needless to say, not all is prosperity and optimism in this field. Shops and stalls must have official licences and should pay tax; there is competition for supplies, customers and premises; officials may prey upon businesses with demands for fees or donations. Tax evasion is said to be routine, perhaps to the general extent of two-thirds. In many fields, such as electrical work, skills remain sadly scarce.

Urban Planning

Few Chinese cities had any experience of urban planning before 1949. The planning introduced under Soviet guidance in the 1950s tended to monumental schemes of redevelopment, often involving the loss of historic buildings such as city walls – Xian is the only large city which retains most of its walls today. Left-wing phases such as the Great Leap and the Cultural Revolution were distinguished by a paucity of new development, and also by indifference to the planning of development. The main building phases have been 1953–7 (under the first five-year plan), the early 1960s and the longer phase since 1979. In all these phases, the main forms of growth have been of the same two kinds – blocks of flats for families, and factories of various kinds.

The total Chinese population doubled, from 541 to 1081 millions, between 1949 and 1987, but between 1949 and 1983 (the year before the definition of 'urban' was changed) the urban population more than quadrupled (*CSY* 1990, 81) in spite of much anti-urban rhetoric. The built-up area of the cities has increased still more. Since 1980 it has been orthodox that growth in large cities should be strictly limited, that medium cities should be encouraged in reasonable development, and that small cities should be actively developed (Kirkby 1985, 207). However, there have been other forces than orthodoxy at work since 1980, and urban growth as a whole has been unexpectedly rapid, as will be shown in the next section.

Until after 1979, planning remained very weak; typically the professional planners and their analogues in the universities and research institutes have occupied a quite narrow decision-space between the capacity of senior officials to make arbitrary planning decisions and allocations on the one hand, and their own lack of funds or authority to make significant positive changes on the other. As a result, predictably, literature in the planning field tends to centre on geometrical layouts and amenity provision in prestige housing developments. Even if there is now more activity at the planning level than before 1979, it is not clear that most of it is more purposeful; 'the decision-makers are not the planners', as one official remarked. Planners still tend to be occupied with prestige projects of the kinds shown to foreigners, rather than with central issues. Senior officials still call the shots in Chinese urban development. Partly as a result of this, Chinese urban land use bears all the marks of non-planning, with fragmentation, inconsistency and widespread inconvenience, particularly for industry (Hodder 1990, 491, 492). Large-scale redevelopment schemes exist, but the bulk of the tremendous physical expansion since 1979 is in the city and town peripheries.

Planned developments in some cities demand special mention, for instance the creation of a sort of mini Hong Kong in the border city of Shenzhen. The most important current planned urban development is Pudong at Shanghai. Shanghai city is located not on the Yangzi, but on the west bank of a minor though substantial tidal tributary, the Huangpu. Pudong is the area to the east of the Huangpu, between it and the open sea, about 300 sq. km in extent. It already has a population of 1.1 million and some 2000 industrial enterprises, and it is now scheduled for modern development – a new port, new industrial installations, upgraded sewage system, metro railway, a bridge over the Huangpu, and in effect a new city (figure 7.1). At present capital is scarce, due to the national credit squeeze, the tendency of national and Hong Kong capital to prefer Guangdong, and international nervousness about China's political future. At the same time the state and Shanghai authorities are anxious for rapid progress in Pudong, to revitalize Shanghai as the nation's future industrial base and the heart of a thriving Yangzi–east China region. Pudong is one place that is openly drawing on rural capital for investment. Meanwhile there is ongoing discussion of the functions of Pudong development. China and Shanghai are already rich in conventional industry. Some people are thinking in consequence of a new Shanghai-Pudong, which should have a much stronger business side than the old Shanghai, and which will be better placed for that reason to challenge Hong Kong after 1997. However, this will depend on both the capacity of Shanghai-Pudong to develop the necessary business mechanisms and techniques, and of course international confidence. Others continue to lay emphasis on upgrading Shanghai's manufacturing base, especially the export side. One group writes, 'In the 1990s Shanghai will change from the country's largest comprehensive manufacturing city to an integrated, multiple-function, key economic city' (Wu Fumi and Li Zhiyong 1990). It is hard to avoid the conviction that behind some of this development is anxiety about competition with Hong Kong after 1997.

In addition to the Pudong development, Shanghai also has three Economic Development Zones, at Minhang, Hongqiao and Caohejing (figure 7.1). All are intended to attract foreign investment, to strengthen Chinese technical expertise and to produce exports. The infrastructure in these zones is said to be varied and reliable. Foreign investors are from Japan, Hong Kong, North America and Europe. Some have obtained land-use rights for long periods in the future. Total projected foreign investment in the three EDZs in 1990 was around 1000 million US dollars (ibid.). Key industrial projects now in the pipeline include expansion of the Baoshan steelworks and Jinshan petrochemicals plant, and power supplies.

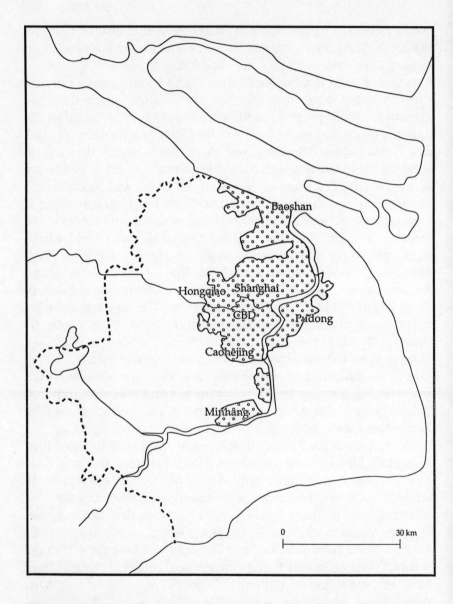

Figure 7.1 Shanghai city and municipality, identifying places mentioned in the text

Urban Housing

One of the most depressing features of Maoist China in its time, considered as a forward-looking socialist country, was the lack of new construction – housing, roads, bridges, railways, even schools. The cities were no exception. Until the post-1979 boom, the great bulk of the urban housing stock was decaying blocks or terraces built before 1949, and often before 1930. Most of this housing was slum material by any standards, though rarely visibly very dirty; and it was always very crowded. Some new housing was built in the 1950s, often in new industrial suburbs of the same period, and some in the early 1960s, also mainly in the suburbs. But housing was in poor shape in 1979, and of course much of the same property is still in use today – slum clearance is exceptionally difficult to put into operation in such crowded conditions. Virtually all urban property, including privately owned housing, was taken into public ownership without compensation during the Cultural Revolution if not earlier, normally by the municipality. At that time its use could be reallocated. However, it has usually been possible since 1979 to enter into negotiations for the return of these houses to their original owners. Anecdotal evidence is that some cities will return a property without a fee; others charge fees (which have to be negotiated); and yet others are not prepared to return properties at all.

In older Chinese housing, there is now a truly awful backlog of obsolescence, the result of thirty years of total neglect. According to a report of 1980, 50 per cent of housing received no maintenance at all. Part of the problem has been families' occupancy of properties as 'welfare benefits' at extremely low rents – rents quite inadequate to provide even for maintenance (Howe 1968). This kind of property is still the usual form near city centres and in the older suburbs; it remains so partly because it is too crowded and too expensive to clear. It usually includes Western-style tenements in two, three or four floors in the main thoroughfares, but also low-built traditional housing in alleys, in back-to-back form, or in old courtyards. Many families in property of this kind have to share kitchens and use public toilets. Socialist principle under the

Maoists did not include any determined attempt to improve urban housing for the working class. Much more has been done since 1979 (Dwyer 1986).

New housing in China generally takes the form of apartment blocks. Since 1979 it has become usual to adopt international construction methods using reinforced concrete; high-rise buildings have very rapidly become part of the skyline of many cities, and even prosperous towns like Shenzhen. Densities of occupation are high by Western standards, though less so by Asian, and according to official figures they have fallen very creditably since 1979 – living space per registered individual was 8.5 sq. m in 1987 against 5.3 sq. m in 1981. The 'households with no rooms' and 'crowded households' figures have also fallen dramatically since 1981, from 28 per cent to 8 per cent taken together (*SYC* 1988, 746). These are impressive achievements, based on a dramatic increase in funding of urban investment. But the 'welfare benefit' model is no longer orthodox; housing is now to be on a commercial basis. New housing can be paid for in various ways: sale to those who can afford to buy; raising rents in company property, perhaps with the help of housing subsidies (many Chinese employees, as elsewhere in Asia, live in housing provided by major employers such as the buses or the public utilities); leasing of a house for a period of years; cooperative systems akin to those of early building societies in England; and so forth. Some of these systems can also apply to old houses.

Needless to say, the quantity of new housing is only part of the story. Quality also has to be considered, and so does infrastructure. Infrastructure in particular is often weak, due to a failure to provide for new building, and very decrepit older installations everywhere, for instance for sewage and water supply. Fittings like locks and hinges, installations like water and drainage pipes, electricity and gas may all take much longer to appear than walls and floors.

Small towns

According to recent details (Ma Rong 1990), China has some 11,500 designated towns (*zhen*), plus about 50,000 rural towns ('townships'). It is sometimes argued that these towns (together with the around 2000 county towns) can be nuclei for the widespread development of commerce and industry in the vast Chinese countrysides, and also of modernization, both technical and social. These are sensible aspirations, but of course they have to be tempered by recognition of the obvious difficulties – weak infrastructures, scarcities of enterprise and resource, limited horizons and official interference with business. In the 1950s and 1960s, due to the nationalization of business, the direct collection of state purchases of grain quotas from the peasants and the suppression of private business, local towns lost both their market and their service functions. Perhaps as many as 280,000 local markets fell out of use in this phase (Ma Rong 1990, 29). After 1979, rural trade began to revive in the new political atmosphere, local industries were encouraged once again, and many small towns began to resume their functions, especially in the most prosperous parts of the country and those with the best traditions of enterprise. By 1982 occupations in towns were recorded as shown in table 7.1.

It is impressive that the job structure of towns gives them a higher degree of dependence on both official and business employment than the cities. That this is paralleled by a lower dependence on agriculture reflects the fact that the boundaries of some of the cities are drawn very widely.

In fact the town figures are quite close to those for the cities; there are much bigger differences between the town figures and

Table 7.1   Kinds of employment in China's towns, cities and villages, 1982 (percentages)

	Towns	Cities	Villages
Industry	37	42	6
Agriculture	21	24	88
Business, Finance	13	8	2
Education, etc.	8	7	2
Transport and communications	7	6	1
Officials	7	4	1
Construction	5	7	1
Services	2	2	negligible

Source: Ma Rong 1990, 30–1, quoting 1982 Census data. Figures have been rounded.

those for the villages, with 88 per cent of workers in agriculture and less than 1 per cent of officials. Interestingly the largest single category of employment in towns is industry – the importance of local industry (township enterprises) has been outlined in chapter 5, and local rural industry or services is the state's preferred option for rural workers who find themselves surplus to need on the land. Ideally such people should take industrial jobs in small towns, living perhaps in hostels, while their families, grain entitlement and registered address remain in the countryside. A survey of 1986 is quoted as showing that 70 million people had in this way 'left the land without leaving the countryside'. Others commute – working in small local towns while returning each night to their village. Usually commuting distances are limited – less than 6 kilometres (Ma Rong 1990, 31). These people usually use bicycles to get to work.

It is still the case that little is said in the Chinese literature on towns on such topics as the service economy and the contribution made by individual businesses – *getihu* as the officials call them. These businesses are still relatively scarce, even in prosperous towns with strong commercial traditions like those of the Pearl River delta in Guangdong. In 1957 and 1958 individual craftsmen and service workers everywhere were roped in to form collectives; such people are now too old to start up business afresh, and their sons are likely to be working for state industries. Painful recollections of direct and personal attacks on such 'tails of capitalism' in Maoist times, led by communist party personnel but often involving neighbours and other local people, inevitably still hold back potential entrepreneurs, especially those with the most relevant family experience. So does the general perception that it is unwise to start up in business without a relationship with an official of some kind, and the recognition that the local officials, well within the limits of their discretion, can make or break such a business at will, by patronage or interference. *Getihu* do not enjoy the position of automatic privilege enjoyed by other township industry among the local officials who are responsible for it and whose official (if not personal) pockets stand to profit by its success.

As the figures in table 7.1 show, small towns are centres of local business in the countryside. Local business has some broad strengths and weaknesses which are worth mentioning. Its most important strength is assuredly the universal scarcity of consumer items in the rural economy, together with consumer services like haircuts, photography, electrical servicing, dressmaking and so forth, particularly if these are to display much quality. Transport is also scarce and profitable: a single minibus (possibly held on contract from a public authority like a town) may easily make a profit around 100 *yuan* per day (around £10 sterling) when all expenses are met; a big bus more than ten times as much. There is now plenty of money among potential customers in most parts of the countryside. The weaknesses of local business are scarcities of supplies, skill, experience and capital, nervousness about the officials on such subjects as taxation and the taking on of employees, and the problem of competition. Business of this kind is of course much stronger in the more sophisticated regions like the Pearl River delta – there are big differences between an area like the delta and the much more backward interior even of a coastal province like Guangdong.

There was little building in China between 1958 and 1978, and the small towns were no exception. Physically, most small towns are quite simple, with older traditional houses and a few blocks of newer flats. Factories or schools may occupy former ancestral temples or halls of prominent families, which are often the only buildings of note. In advanced parts of the country, there is now often a good deal of new building, particularly small factory buildings in which floors can be let separately to various enterprises, and in prosperous towns there are also new terraces of shops with family accommodation above, occupied by *getihu* businesses. Infrastructures (street maintenance, water supplies, sewage, electricity supplies) may be both limited in scale and very run down. There is a virtually limitless job of modernization to do, for instance in transport and its various ancillary trades like motor vehicle repairs, in electrical work of all kinds, in the public utilities and the banks, in training for youngsters. Everything needs to be done at once. Business trust and experience, business and technical competence, the fitting up of premises, the reconstruction and modernization of social habits – all these take time, and progress is bound to be discontinuous.

Rural–Urban Relationships

In 1984, the central committee of the Chinese communist party agreed on a range of measures intended to strengthen and develop urban reform (An Zhiwen 1985). The essence of these was the replacement of the Maoist 'closed, vertical' structures of authority and management by 'open, lateral' structures. This was to be done by the abandonment of various aspects of the command economy and their replacement by voluntary, business-based relationships using commercial considerations. 'The urban economy as a whole, which had been dominated by the product economy ignoring the laws of value, is now turning into a planned commodity economy' (ibid. K9). In addition, the committee set its seal on the policy of greatly widening the administrative authority of the cities, as against the prefectures which are the ordinary sub-units of the province. About one-quarter of Chinese counties were already being transferred to the administrative responsibility of cities. The same policy has been continued in subsequent years; at the present time about 35 per cent of counties, with probably around 45 per cent of rural output, are under the 'leadership' of cities. In Liaoning and Jiangsu, both advanced provinces, all counties are now 'led' by cities. Here too the intention of policy was to strengthen the commercial, non-official side of Chinese organization; not for the first or last time, the communist party imagined that action taken in the bureaucracy could be relied upon to stimulate change on the ground. Cities are advised that the purpose of the administrative reform is to enliven local economies and to strengthen and facilitate healthy business relationships, not to enable cities to bully the rural counties or lay hands on their funds. The heart of the matter appears to be the party's broad policy for the shedding of rural labour by the agricultural sector, and the integration of the people displaced into industrial and service occupations. Consistent with this is the planned movement of various urban industries, or parts of industries, to the countryside, and the further development of direct supply-and-demand business relationships between city and countryside (*People's Daily* 5 January 1985). Developments of this kind are often singled out for warm praise in the press. Objectively, however, it is not clear that very much has changed at the official level. City-based officials have taken the place of county-based or province-based. Local officials may now be more important than formerly, but the power of official individuals and the strength of official culture are not touched. Realistically, it is not possible, and would not be even if information were better, to analyse the effects of changes of this kind. In many urban peripheries there has been significant economic development, but much of this might well have taken place without the administrative changes – should have done so, if the administrative system could have been relied upon to be 'neutral' in accepting business-based developments.

The Strength of the Cities

There has been much in both the tone and the substance of official pronouncements since 1949 to indicate that the communist party seeks the welfare of the Chinese countryside at least as earnestly as that of the cities. In terms of practical management, however, the cities have done very well out of communist party rule, partly (it would seem) out of necessity, partly out of inadvertence and partly as a spin-off from the party's insistence on industrial growth, of which the greater part has been located in cities. One consequence of urban privilege has been the haloes of rural prosperity which surround Chinese cities, and which are one of the main forms of local differentiation in the countryside. These haloes are areas from which day-to-day marketing by local peasants in the cities is feasible; a typical radius in densely populated areas is two or three kilometres. Urban markets generate prosperity because prices are high and generally rising, and urban demand is able to pay prices which are perceived to be high, partly because urban customers pay cash to country people who (like most peasants) do not otherwise handle much cash, and partly because urban demand is fortified by urban incomes which are by rural standards quite high. Urban and rural lifestyles differ so much that direct comparisons cannot be made, but in terms of cash, the 1211 *yuan* per annum 'living expenditure' per person in the average urban household in 1989 cannot but compare handsomely with the 535 *yuan* in the average peasant household (*CSY* 1990, 279, 298) – the rural figure is less than half the urban. After agriculture, industry is the largest employer in China, with 96 million workers, of whom 43 million work in state-owned units and 18 million in collective-owned (*CSY* 1990, 106, 107). Incomes per worker in every other part of the economy are higher than in farming, usually by about 50 per cent, and this difference has widened since 1978 (*CSY* 1990, 134). Most workers in the non-farm parts of the economy (industry, building, transport, commerce and so forth) are of course urban-based.

The industrialization programme is the heart of urban prosperity, as of urbanization itself. It has been shown (chapter 6) that the industrialization programme has run with remarkable consistency through the various policy revolutions of the past forty years, and that it is still in full force in the present phase. In China as in most countries, industrial enterprise is primarily an urban phenomenon. Typically, the urban 30 per cent of the population produces 70 per cent of the industrial production. In Chinese industry as a whole, 11 per cent of the work-force produces 47 per cent of the national product (*CSY* 1990, 26, 107) and around 70 per cent of this achievement is made in the 450 cities, only 174 of which have more than 200,000 inhabitants. We may turn to cities of more than 200,000 population (*CSY* 1990, 63) to reveal a similar situation: these 174 cities, with 16 per cent of Chinese population (but no figure for work-force) produce 50 per cent of industrial output, and hence around one-quarter of total national product.

In addition to higher incomes urban citizens also enjoy price subsidies for rural outputs when these are sold in the cities. These subsidies have increased greatly under the reforms since 1978, apparently because the state has been reluctant to allow purchase price increases towards farmers to be reflected in urban sales prices. State subsidies have increased from 1.1 thousand million *yuan* in 1978 to 35 thousand million *yuan* in 1989, more than one-tenth of state revenue (*CSY* 1990, 209, 224) – and this figure can be doubled by other kinds of subsidies, such as provincial subsidies and concealed subsidies like tax incentives. At least 120 commodities are involved, from grain and firewood to beer and silk, but most of them are also available, unsubsidized, on the open market. Price subsidies would appear to account for 10–15 per cent of urban households' incomes; more for poor people. Additional subsidies in fields like housing are said to bring the figure to 50 per cent (some critics say 80 per cent) (Zhang Ping 1990, 31). Subsidy on this scale is self-evidently very expensive, and of course much of it (especially price subsidy) is indiscriminate – it helps out well-off families as well as the poor. It obscures the true cost of labour and adds a further level of distortion to the price system. It also self-evidently represents favour to the cities – few such subsidies are available to rural people.

Even consumer subsidies are not the end of the story. Many urban enterprises, especially state industries, enjoy supplies of coal, water and electricity at low state prices. These items are provided (cheaply) by rural areas. Urban primary schools are free; rural families have school fees to pay. Wages are paid in cash and rents are extremely low in the cities. In general, the official price system is designed to keep down the cost of state purchases. This inevitably favours the city, where these purchases are generally used.

Rural land is often lost to villages by conversion to urban use (waterworks, airports, industrial plant, suburban growth). In these cases the state may simply transfer the village land (which belongs to the village either directly or through the local town) to the city, after which it belongs to the state. Inevitably the city is the gainer.

Rural savings are notoriously used by the banks to finance urban developments. The price 'scissors', the relationship between agricultural and industrial prices, have tended to move against agriculture since 1985 (Aubert 1990, 28). Attention has already been drawn to the long list of official charges and fees which are levied upon the peasants. Most of these do not exist in the cities. In addition, in China as in other Asian countries, cities enjoy various social advantages in addition to high incomes – much more variety and stimulus in lifestyle, more opportunity for teenagers, more entertainment, and of course access to factory work which has fixed hours and is generally clean, unlike work on the land.

'Urban bias' is a condition identified by Lipton (1977) and others, mainly in Third World conditions, in which the state is believed to influence the allocation of incomes so as to favour urban communities to the disadvantage of rural. Influenced no doubt by official (particularly Maoist) rhetoric, Western writers have tended to suppose the Chinese systems to be relatively free from this particular corruption; but the truth is that it has been routinely practised through all the political phases since 1949, and is still practised. In part this has resulted from the imperatives of the industrialization policy; but in part (especially where consumer subsidies are concerned) it seems to result from a tenderness towards the interests of the urban populations which is not at all consistent with official rhetoric.

Commerce and the Service Economy

The Maoist background

Maoist theorists argued in the 1960s and 1970s that spontaneous business of any kind carried the seeds of fresh growth of capitalism, and was hence to be suppressed. In various phases between 1958 and 1976 serious attempts were made to break up private trading networks and stifle trade among individuals, and even among village cooperatives. On the whole these efforts were successful, at least for periods of years. In China as a whole, 73 per cent of retail sales outlets were closed between 1957 and 1979. Numbers of urban retail outlets fell even more sharply, by fully 90 per cent in the same phase (Leeming 1985a, 24). Wholesale trade had meanwhile been effectively nationalized during the 1950s. During this long phase, trade was not supposed to take place except through organs authorized by the state. Understandably, in these conditions supplies of all kinds became scarce, even supplies of the simplest vegetables in the cities. Understandably too, the state-authorized organs of trade proved inadequate to supply consumers other than those with state-authorized priorities, such as state factories and canteens, and the exporting organizations. Urban shops and restaurants (even those run by the urban authorities) and the rural supply and marketing cooperatives, all found themselves very short of supplies – sometimes disastrously so. Finally, service trades like photography, dressmaking, haircutting and electrical repairs practically disappeared in this period. In 1978, for the 5 million urban people of Beijing, there were only 259 individual businesses (Chen Jian 1985).

These policies have been reversed since 1979. The communist party and central authorities are now strongly supportive, in principle, of local enterprise in trade, transportation, services, catering and business of all kinds. Businesses may now be run collectively by local units (county, town, village), or by individuals or cooperative groups. However, it appears that at the grassroots the habits of regulation are proving difficult for the officials to shake off in their relations with private business, while at the same time the old tradition of official economic activity proves tempting. In short, the new official policy is often inhibited in practice by the old official outlooks.

The private small-scale economy

In the China of the present, individuals or cooperative groups can set up their own businesses provided they have a license. The license must be sought at the local office concerned with such matters, and includes permission to locate in a certain place. These are the *getihu* ('individual household businesses') which have already been mentioned. They are usually run by whole families. They may occupy old buildings such as ancestral temples (the only big buildings in most traditional towns), or even sheds on the edge of town. But in addition many towns, and even villages, are now building premises for rent by small businesses, usually in the old terrace form used earlier this century, with open shops or workshops on the ground floor and living premises behind and above. These terraces of shops are also being built in the cities, especially in the new good-quality suburban neighbourhoods, but in the cities *getihu* most often occupy old premises.

Small private businesses may be general shops, small cafes, service businesses such as hairdressers or photographers, or small manufacturers such as dressmakers. Some businesses started out in the free markets eight or ten years ago, and have now graduated to fixed premises. They depend totally upon the goodwill of the officials, in granting a license and in accepting the business's existence month by month. Below the level of these businesses in fixed premises are itinerant noodle or bun sellers, who sell from an old-fashioned handcart, and also the groups of cobblers and others who occupy particular street corners in the cities; but all these people need licenses and, of course, toleration. The officials themselves need such businesses, however, to generate revenue through licence fees and taxation, and to stimulate business in competition with other local towns.

Needless to say, problems arise. A body of literature about the *getihu* has accumulated, some of which has been translated into English (*Chinese Economic Studies* 1987–8). Many problems relate to relationships with the officials, who may persecute *getihu* with unreasonable demands and even force them out of business. Grey areas, such as sites, hiring of workers or competition with official enterprises (as with buses), all invite interference. Taxation is always open to negotiation, but negotiation does not necessarily satisfy either side. *Getihu* are recorded in many surveys as fearful of major changes of policy, which might cast them once again as 'tails of capitalism' and social outcasts, as during the Cultural Revolution. But incomes of *getihu* families who work hard, often for long hours, can be substantial – much higher than those of junior officials. People with natural entrepreneurial qualities can make very sound livings; and as some writers point out, the public service performed by these people and the taxes they pay are produced without any support at all from the state.

The official commercial economy

Vivienne Shue (1990) has recorded two impressive examples of official enterprise at county level, Shulu in southern Hebei and Guanghan in Sichuan – both quite close, as it happens, to major cities. In Guanghan the officials are in the forefront of direct investment and entrepreneurship, with a convention centre and hotel in the county capital, to include shops, bars and restaurants. In Shulu the parallel development is a shopping centre accommodating local businesses, together with an open-air market. Both depended upon the privileged position of the officials in raising capital, and no doubt on their special privileges in other ways, such as getting building work done. Both have done much to focus and stimulate improvement in the two towns. Both, of course, represent the continued involvement of officials in business, even in a phase in which official intervention is supposed to be withdrawing; but both also represent an up-to-date and entrepreneurial outlook which is new in official intervention. The same is true of town and village factories which belong to the local administrative units and are managed by official appointees. In all these kinds of business, local official enterprise fills public coffers (through charges and fees), and gives local officials a record of achievement which is consistent with public policy, and money to handle which lies outside official budgets. It is not suggested that such money is usually corruptly managed, though no doubt that can happen; but even without corruption the handling of the considerable profits that may be generated by business can transform the outlook of local officials very quickly.

Migration and the New Urbanization

In 1984 the communist party relaxed its long-standing hostility to the migration of peasants away from the land. This was in the context of a growing recognition of the scale of the surplus labour problem in the countryside. By 1985 and 1986 it was being argued that around one-third of agricultural labour was surplus to need on the land, and that in view of the rapid growth in the labour supply during the 1980s and 1990s, the countryside could usefully shed between 200 and 250 million workers in the decade between 1985 and 1995 (*Nongye jingji wenti* 1986; Lei Xilu 1990). This immense number represents about half of the rural workforce. Migrant workers in China are supposed to 'leave the land but not leave the countryside'; they are expected to take up jobs in township enterprises or start up *getihu* businesses, all preferably in their own localities.

In fact the township enterprise sector has shown a remarkable capacity for expansion, raising its employment from 60 million people in 1986 to around 95 million in 1989. As suggested in the discussion of state industry, this capacity for expansion no doubt owes something to the rigidities experienced by the latter. Nevertheless, even this expansion has not been sufficient to mop up the surplus of rural workers. In recent years most cities, particularly the biggest, have experienced very extensive immigration of country people, and the same is true of even modest rural towns in the most prosperous parts of the country, such as southern Jiangsu around Shanghai and the Pearl River delta in Guangdong.

The numbers themselves are impressive. An article in the official *People's Daily* (Shu Yu 1989) proposed a figure of around 50 million migrants, one in twenty of the whole population, and around one in ten of the total labour force. Of these people, some 10 million were said to be living in China's 23 cities with more than 1 million people – for instance, 1.8 million in Shanghai and 1.1 million in each of Beijing and Guangzhou. These suggest high proportions of migrants in the great cities, over 20 per cent in Shanghai and Beijing, and 40 per cent in Guangzhou. A number of studies of migrants have been published in China, and (perhaps understandably in view of China's vast size and variety) not all of them are consistent. Some have found that the majority of migrants are peasants, others that the majority are either townspeople or peasants who have lived in local towns at home. Most migrants come either from the local countryside, it appears, or from the vast hinterland of the 'Centre' belt of provinces. The destinations which exercise most attraction are in the East region. But even if at least 20 per cent of migrants make for the great cities, the great majority, evidently, do not; and there are few studies of these. Women migrate as well as men, looking for factory jobs as a rule. Not all migrants are young; many are of an age to be heads of farming households. Some studies made in China suggest that working peasants have often become disenchanted with farming and with the life of the countryside for some of the reasons mentioned in chapter 5, and have decided to make a dash for an alternative life in the cities (Wu Huailian 1989). It is important to recall that for twenty-five years (from 1960 to 1985) it was very difficult, if not impossible, for a rural family or even individual to go to live in town. Both pent-up ambition and the perception of an exceptional chance are certainly involved in the present migration. Urban bias, widely perceived in the community, must be a powerful incentive.

Migrants take jobs in factories (especially township and other unofficial enterprises), building, commerce, the service trades and domestic service. Inevitably they put pressure on the city utilities, such as water and electricity supplies, and also on food supplies and the urban buses. They provide cover, unwittingly, for criminals on the run. They occupy an important place in the present proliferation of low-grade urban services, which have been badly missed in the Mao and post-Mao years, and hence are one of the several factors in the reconstruction of the crowded, dense urban landscapes of China before the Maoists, and their closer approximation to urban landscapes elsewhere in Asia.

EIGHT

Features of Regional Change

Writers on China often give particular prominence to the distribution of poverty – more than they would, perhaps, to the same topic in North America or India. This appears to be justified by China's 'socialist' claims and the management of the country as a unitary state; but as has been shown, the communist party has a wide range of preoccupations other than egalitarianism; and even in Mao's time, when equality was pursued at considerable cost in suppression of growth, the environments took a powerful hand in resisting it. In reality it is less than realistic to look at the regional distribution of prosperity without due reference to the regional production systems which underlie it. In this chapter it is intended first to review materials which relate directly to prosperity and poverty, and second to look briefly at some regional types in development.

Prosperity and Poverty – Regional Patterns

The best single document for the geography of Chinese poverty is the map of 'Rural poverty areas (at *xian*, county level) receiving national support during 1986–90', designed by Johannes Küchler (1990) and reproduced here by permission (figure 8.1). The Chinese data used adopt several different levels of definition of rural poverty for different parts of China, but 'the common denominator . . . was that at least 40 per cent of the total rural *xian* (county) population had an annual per capita net income below 150 yuan' (Küchler 1990, 131). This would be, by direct conversion which of course is not realistic, around £17 sterling – a more realistic notion of equivalence might be around £60 sterling. In fact the map is composite, and includes areas (which are distinguished by differing symbols) of three kinds. The third kind, defined by rainfall below 400 mm and involving only parts of Gansu and Ningxia in the centre-north, is to a degree anomalous; Chinese atlas maps show that practically all of Inner Mongolia, Gansu and Xinjiang, and most of Qinghai, have less than 400 mm precipitation; but of course in most of these territories population is very scanty compared (for instance) with Ningxia.

'Poverty counties' in figure 8.1 appear to be of three kinds. First, scattered very widely, and including some counties even in prosperous provinces like Shandong and Guangdong, are those in the mountains. Mountain environments, particularly in the south, are not necessarily poor, as will be shown in a later section, but they are often inaccessible, have low technical and sometimes low resource levels, and are usually backward. The same may be said of isolated localities in the Pyrenees or Appalachians. It is noticeable that most of these counties lie on the borders of provinces. These provincial borders are often mountainous; but, in addition, there is probably a tendency for counties out of sight of the provincial capital to be equally out of mind, and in this respect the Maoists differed very little from the reformers of the present.

A second group comprises counties in Xinjiang, mostly along or close to the frontiers with the Soviet Union. These counties occupy difficult arid and usually mountainous environments, though not more so than others further east in Xinjiang or Gansu. They may perhaps enjoy inclusion in the relief programmes represented in the map as a consequence of frontier status.

A third group is in some ways comparable, but lies in the zone of weak, but relatively densely populated, environments along the southern border of Inner Mongolia, in that province itself, in Shaanxi, Gansu and Ningxia. This is the southern fringe of the Gobi desert, most of which lies in the Mongolian People's Republic ('Outer Mongolia') to the north. It is a zone of exceptional weakness in China.

It is worth noting that in the whole north-east there are only two qualifying counties, both in a Mongol nationality area, and that in

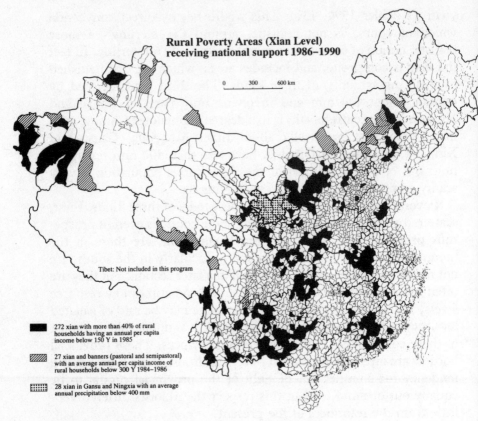

**Rural Poverty Areas (Xian Level)
receiving national support 1986–1990**

0   300   600 km

Tibet: Not included in this program

272 xian with more than 40% of rural
households having an annual per capita
income below 150 Y in 1985

27 xian and banners (pastoral and semipastoral)
with an average annual per capita income of
rural households below 300 Y 1984–1986

28 xian in Gansu and Ningxia with an average
annual precipitation below 400 mm

Figure 8.1   Rural poverty areas, 1956–1990
Poverty areas of several different kinds are shown in this map, as
the key indicates. Nearly all share two qualities – isolation and
difficult environments. 'Xian' are counties.

Source: Küchler 1990, 130. Map designed by Johannes Küchler and drawn by
Wolfgang Strands, ILO, Berlin. Reprinted from Jorgen Delman, Clemens Stubbe
Østergaard and Flemming Christiansen (eds), *Remaking Peasant China* (1990), by
permission of Aarhus University Press and Professor Küchler.

the bulk of eastern China only a few provinces have more than a
handful. The map is constructed on the basis of 'national support'
programmes; it would appear possible that trouble was taken to
keep eligibility within reasonable bounds, possibly with the help of
a quota system.

Study of poverty suggests in turn study of prosperity, and also of changes over time. Table 8.1 gives figures by sample provinces and regions for gross national product per person, and in figure 8.2 all provinces are mapped. The three great cities have far and away the highest figures, followed by what are generally agreed to be the three most advanced provinces – Liaoning, Guangdong and Jiangsu, all coastal. All coastal provinces, except Hebei (which narrowly fails to qualify) and Guangxi in the far south, have figures above average, and their mean is also well above average. Below average are four great blocks of territory with aggregate population of 626 million, 56 per cent of China's total – the north China plain, the middle Yangzi, the south-west and the centre-north (table 8.1). All interior provinces taken together have an average of 1059, 76 per cent lower than the coastal average.

Figures for gross national product per person are not available except for very recent years, but those for gross output value of

Table 8.1 Gross national product per capita, 1989, by selected provinces and regions

Province/region	GNP per capita (yuan)
Beijing, Tianjin, Shanghai	4500
Liaoning	2359
Guangdong	2183
Jiangsu	1892
All coastal provinces, plus Beijing	1860
China average	*1388*
Hebei, Henan, Shangdong (north China plain)	1247
Hubei, Hunan, Jiangxi (middle Yangzi)	1133
Shaanxi, Gansu, Ningxia, Inner Mongolia	1075
All interior provinces	1059
Sichuan, Guizhou, Yunnan	891

Source: CSY 1990, 35

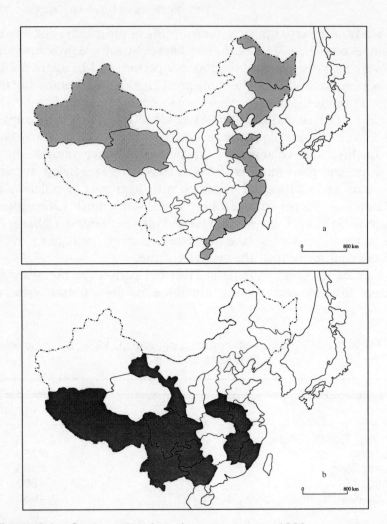

Figure 8.2 Gross national product per person, 1989
(a) Provinces with levels of GNP per person above the national
average. High figures for the coastal provinces and the north-east
relate to high levels of development. Those for Xinjiang and Qinghai
appear to relate to state spending, in very large areas with scanty
populations. (b) Provinces with levels of GNP per person below
75 per cent of the national average. Tibet and (to a lesser extent)
Gansu are poor provinces in most respects, and those of the
south-west such as Yunnan are low in development activity. To a
much lesser extent the same must be said of Henan, Anhui and
Jiangxi, in the central belt.
Source: *CSY* 1990, 35, 83

agriculture and industry are available by province from 1982 (*SYC* 1983, 21). The results are shown in figure 8.3. Coastal prosperity is much less evident, though the three great cities are already paramount. High values are displayed by the north-east, Jiangsu and Zhejiang, and Hubei. Shandong narrowly fails to qualify, but Guangdong and Fujian have low values. The centre-north falls into the lower category, along with the south-west. Particularly interesting is the group of poor provinces inland in eastern China – Henan, Anhui, Jiangxi – which still rank as poor in the map for 1989. The differences between the two maps are hard to rationalize with confidence (in part of course because the figures are on two different bases), but the distribution of 1989 insists upon the primacy of the coastal provinces, as might be expected. The difference in status of the centre-north, including Inner Mongolia, is more apparent than real; both Inner Mongolia and Shaanxi are marginal in the 1982 map.

In figures 8.4 to 8.6 an attempt is made to analyse changes since 1982 a little more exactly, with the help of separate statistical series for 1982 and 1989 for agricultural outputs, industrial outputs and values of retail trade. The period covered by these figures is not long, and the 29 provinces (not including Hainan, whose figures are aggregated with those of Guangdong) represent a very coarse network, but some features do emerge in the maps.

There is rather less agreement among the groups of provinces shown in the maps than might have been expected, though for all three indicators there are few provinces which lag far behind the average, and, except for industry, only a few which rise above it – sometimes very much above. For agriculture (figure 8.4) improvement includes some provinces with marked advantages (Guangdong, Beijing), but more which are notoriously backward (Gansu, Guizhou, Anhui). Among the relative laggards in agriculture are some provinces with grain-base status, such as Jilin and Hunan, and others for whom the 1982 level was already high (Shandong, Jiangsu, Shanghai). The retail sales maps (figure 8.5) seem to follow 'community' rationales, perhaps understandably. Prominent provinces are those on the coast southwards of Jiangsu, together with Beijing and Tianjin; provinces falling below the average seem to be in most cases those with vast distances, difficult conditions and limited retail networks (Inner Mongolia, Xinjiang,

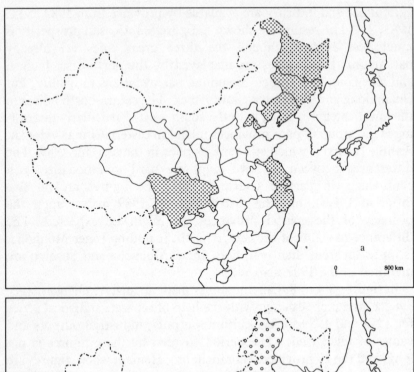

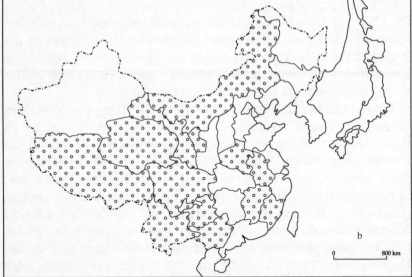

Figure 8.3 Gross output values of industry and agriculture per person, 1982
(a) Provinces with values above the national average. Compare with those in figure 8.2a. (b) Provinces which have less than 75 per cent of the national average. Compare with figure 8.2b; there are several similar features, notably the Henan–Anhui–Jiangxi group.
Source: SYC 1983, 21, 106

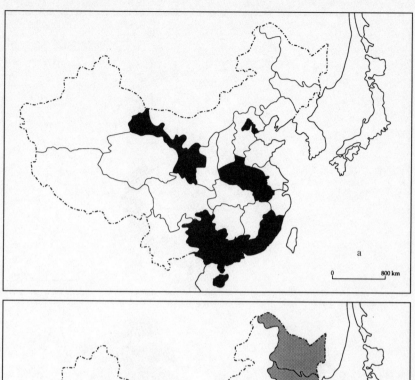

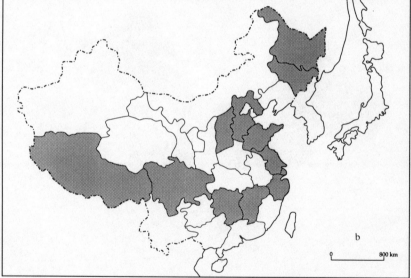

Figure 8.4 Agriculture: growth in gross values, 1982–1989
(a) Provinces with growth in value of agriculture above the national
average. A scattered group of provinces, some of which (Gansu)
started from very low bases. (b) Provinces where growth was less
than 90 per cent of average. For some of these (Jilin, Heilongjiang),
this low result seems to indicate relaxation of previous state
pressure; for others (Sichuan, Jiangxi) continuity in already
favourable conditions; for yet others (Shandong, Jiangsu, Zhejiang)
an energetic response to the new opportunities under the reform
policies.
Sources: *CSY* 1990, 520; *SYC* 1983, 152.
Note: 'Agriculture' does not include township enterprises.

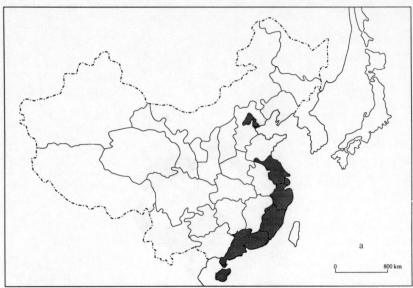

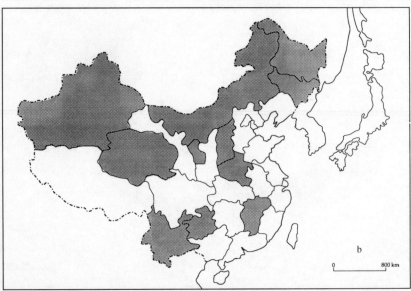

Figure 8.5 Retail sales – growth in gross values, 1982–1989
(a) Provinces with more than average growth in retail sales. The
southern group of coastal provinces, with the three great cities, are
indicated. (b) Provinces with growth in retail sales less than 90 per
cent of the national average. These are mainly provinces with vast
areas and weak retailing systems (Inner Mongolia) or those noted as
relatively backward (Jiangxi).
Source: *CSY* 1990, 591; *SYC* 1983, 374.

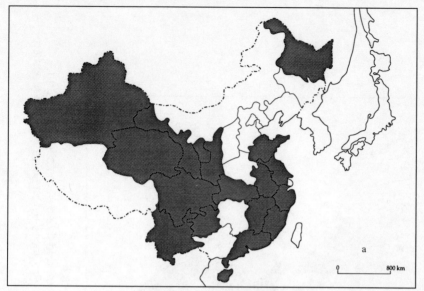

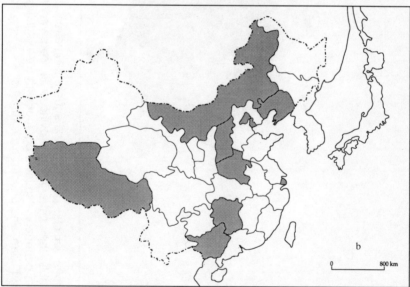

Figure 8.6 Industry: growth in gross values, 1982–1989
(a) Provinces where growth has been greater than the national
average. Most provinces are shown, but not the three great cities or
a group centred on Beijing–Tianjin. (b) Provinces registering less
than 90 per cent of average growth. Here the three great cities and
Liaoning are prominent, together with some provinces which have
very little industry (Tibet, Jiangxi) and some whose industry is
impoverished (Shanxi).

Sources: *CSY* 1990, 396; *SYC* 1983, 226

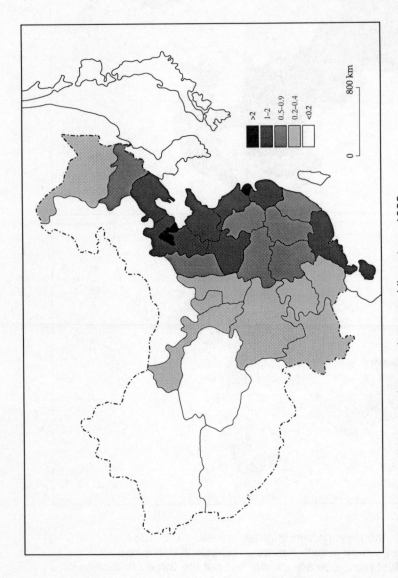

> 2
1–2
0.5–0.9
0.2–0.4
< 0.2

0   800 km

Figure 8.7   Civil motor vehicles per thousand square kilometres, 1989
In this map, predictably, the central Asian provinces all score very low, and those of the south-west and
northern central Asia periphery also low, together with Heilongjiang. Provinces which score high are those on
the coast apart from Fujian, and of course the three cities.

Source: CSY 1990, 432–3

Yunnan), though a few (Jiangxi, Henan) are among those generally considered relatively backward. For industry, the maps are rather different (figure 8.6). Here seventeen provinces (more than half) have figures above average, but most of those with figures very much above average (Shaanxi, Sichuan, Yunnan) have been notoriously backward (but not Guangdong, also with a high figure, though not the highest – that honour goes jointly to Qinghai and Ningxia). Figures noticeably below average include the three great cities (with relatively little scope for further growth, it would appear), Liaoning and – surprisingly – the coal provinces of Shanxi and Henan, which are perhaps hamstrung in the figures by low coal prices.

One more map may conclude this display of evidence for prosperity and poverty, and that is the plot for civil motor vehicles per thousand sq. km (figure 8.7). This is introduced because it offers evidence which comes close to justifying standard prejudices on the distribution of prosperity. Here the three great cities stand alone, with a chain of mainly coastal provinces following, but including Henan and excluding Fujian. Following again are Shanxi, Jilin and the middle Yangzi provinces, including Anhui, with Fujian – the heart of the south. Following are very extensive provinces with either scanty rural populations like Gansu, or backward reputations like Yunnan, and finally the main central Asia provinces like Xinjiang, including Inner Mongolia. According to this indicator, high and low scores are separated by a factor of more than ten – very much more than the rather limited differences which separate the 1982–9 growth rates in agriculture, retail sales and industry.

Attention was drawn earlier to China's investment history, particularly in terms of sectors of the economy. Something may be added here about the regional distribution of investments. Table 8.2 gives details of total fixed investment by a series of important regional groupings, for 1982 and 1989. The effects of the reform policy are clearly visible – coastal provinces now account for 58 per cent of investment, against 48 per cent in 1982, with a corresponding fall in the share of inland provinces. The share of the central Asia provinces has fallen dramatically, and of the three great cities and the north-east, though far less dramatically. The inland south's share has hardly changed – it was already low. A striking improve-

Table 8.2   Investment in fixed assets by regional groups, 1982 and 1989 (percentages)

	1982	1989
Beijing, Tianjin, Shanghai	15	12
Hebei, Henan, Shandong (North china plain)	14	18
North-east	15	12
All coastal provinces plus Beijing	48	58
Inland provinces	52	42
Mid-Yangzi with south-west (6 provinces)	17	16
Xinjiang, Tibet, Qinghai, Gansu, Inner Mongolia, Ningxia (central Asia provinces)	9	5

Sources: For 1982, *SYC* 1983, 335; for 1989, *CSY* 1990, 144

ment has taken place in the share of the north China plain provinces.

Figure 8.8 shows the geography of investment for 1989. The maps are constructed on the basis of the national average figure of 372 yuan per person. Only the three great cities fall into the highest category. In the rest of the 'above average' group are six coastal provinces, Heilongjiang and Shanxi with their oil and coal respectively, and two central Asia provinces with low populations and high state interest. The most marked feature of the 'below average' group is the lowest category, of seven southern provinces. In view of figures of this kind, it is sometimes tempting to conclude that the low-performance inland south of China is a creation of the state, and that continuity in the slow-moving and traditional south owes something to the self-fulfilling prophesies of investment priorities, especially those of the state.

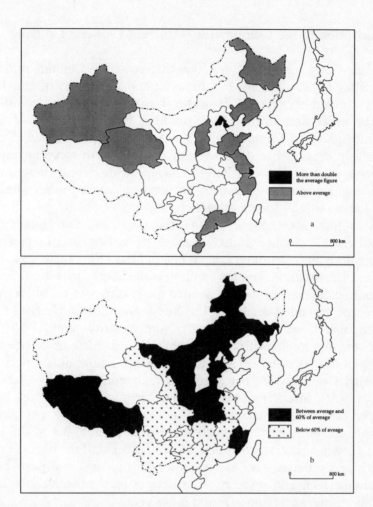

Figure 8.8 Fixed assets per person by province, 1989
(a) Provinces (only the three cities) in which fixed investment is more than double the national average, together with a group of provinces where it is above average. Most of the latter are coastal, but Heilongjiang and Shanxi probably qualify as energy bases. Xinjiang and Qinghai are of special interest to the state. (b) Provinces with low investment at two levels – between average and 60 per cent of average, and below 60 per cent of average. Most of the former are northern provinces; most of the latter southern, including all provinces of the inland south except Hubei, plus Guangxi, which is coastal.
Source: *CSY* 1990, 144, 83

Polarization in the Countryside – Current Chinese Thinking

It is generally recognized in China that power and money inhabit the cities rather than the countryside; and the reality of this has been outlined in chapter 7. However, business is no longer confined to the cities and towns. It is reaching important parts of the countryside, and bringing with it money and opportunity, if not usually power. It has done so with great force in some favoured regions, such as the Pearl River delta, to whose people the growth of Hong Kong owes so much; but even in much more ordinary rural areas similar forces are generating similar results.

It is now accepted, even in official circles, that increasing circulation of money must generate important social changes, particularly in the countryside. While in Mao's time there was very little differentiation among commune members, in recent years economic differences have generated many different social groups. A recent extended analysis (Lu Xueyi and Zhang Houyi 1990) distinguishes: agricultural workers (the majority, around 56 per cent of the total, with various sub-groups such as specialist, prosperous, subsistence and impoverished households); peasant industrial workers (24 per cent); privately employed hired labour (4 per cent); intellectual peasants, i.e. technical workers, teachers (2 per cent); individual workers such as craftsmen (5 per cent); private entrepreneurs, normally employing labour (0.1 to 0.2 per cent); township enterprise managers and other staff members (3 per cent); and officials, also of several kinds (full-time and part-time, local chairmen and secretaries, leaders of local organizations like youth league secretaries; around 6 per cent).

No doubt understandably, the authors of the analysis quoted are less than clear about its policy implications. They (and the communist party) have no difficulty in criticizing exploitation of labour, tax dodging, illegal activities such as currency dealing, and money made through favours from officials. With less assurance, they accept the positive role of individual businesses in furnishing enterprise and catalysing local development possibilities. But even within the law, some rural households are now able to earn tens of thousands of *yuan*, or even hundreds of thousands, while ordinary households earn little more than 1000 *yuan* (Lu Xueyi and Zhang

Houyi 1990, 63). Peasants are dissatisfied about this polarization, but (except where there are illegal components) it cannot be suppressed. Ordinary peasants, even those taking up specialist outputs, are also upset about the 'scissors' of prices for inputs which rise faster than prices for outputs. The argument easily slips out of the problems of polarization and into the problems of the countryside. Other writers point to the general improvement in peasant incomes which has taken place, argue that the increase in differentiation has remedied a previous irrational egalitarianism, and insist that getting rich through hard work is not polarization, but social justice (Huang Daoxia 1987).

Some Regional Examples

It has already been shown that the People's Republic contains a vast variety of resource endowment and human development, whatever its uniformitarian pretensions. There are broad trends of difference among the thirty provinces, and of course an infinity of variation among the more than two thousand counties, and within them. It is hoped with a few examples to develop these notions a little further.

Liaoning and the north-east

China has three main foci of relatively advanced development: Jiangsu and Shanghai, with the adjacent province of Zhejiang; parts of Guangdong, especially the Pearl River delta around Guangzhou; and Liaoning. In this brief review, it is intended to focus first upon Liaoning and Guangdong, prosperous provinces which display contrasts in development experience.

The present population in the three provinces of the north-east (Liaoning, Jilin and Heilongjiang) is 98 million, with nearly 14 million non-farm people in cities of more than 1 million (*CSY* 1990, 83, 84). Eighty or ninety years ago, in a phase of contention among China, Russia and Japan for control of the north-east, its total population was at most a few millions, and it had only one city, Mukden (the present Shenyang, now China's fourth largest city), which then had a population of less than 200,000. The reason for international contention was the north-east's vast resources and relative lack of population – in the age of rampant colonialism, it was prime 'colonial' territory, Asia's share of the world's thinly populated, mineralized temperate grasslands, the Far East's Midwest or New South Wales. The Chinese north-east was the last of these prime colonizable territories to be recognized and developed. It fell into Japanese hands in 1906 and was later developed as the Manchukuo protectorate, but throughout the Japanese phase the Chinese population grew, through labour migration from Shandong and the north China plain; in 1930 its population was already given as 29 million, with an urban population of more than 1 million (Cressey 1934, 218). In some ways, China's modern development began under the Japanese in the north-east, before liberation. But population densities remain relatively light, and both farm and industrial resources relatively plentiful. Modern development has continued since liberation in the best conditions of environment conjoined with population which exist in China. This modern development has taken very much the forms preferred by the communist party; it is perhaps not too much to say that the north-east is China as the communist party would like to see it. Total output (GNP) in the north-east is 11.8 per cent of the China total; but the population is only 8.8 per cent. The birth rate in the north-east is 16.4 per thousand, against the national figure of 20.8. Thirteen per cent of Chinese graduates come from the north-east (*SYC* 1990, 35, 83, 679). A wide variety of such indicators can be found, most of them flattering to the success of official policies in this region.

Liaoning is the foundation of the north-east, as earlier it was the foundation of the Japanese protectorate. It was specially favoured in terms of investment in the First Five-Year Plan and subsequently; it now has 6 per cent of all Chinese investment (against 3.5 per cent of the population), 80 per cent of it state-owned (*CSY* 1990, 144). In a list of 1986, Liaoning had 7 per cent of China's major plants, the same as Shanghai (*CSY* 1988, 353–68). But in recent years there are signs of ageing in the industrial equipment and resources of Liaoning, and also competition from other regions. Liaoning's coal output is now only 5 per cent of China's total. Although steel output remains outstanding (20 per cent) and the Anshan state steelworks remains one of China's most profitable companies, Anshan is said to be in need of extensive fresh investment, which in present conditions cannot be forthcoming. Another problem specified by the governor of Liaoning in a recent interview (Li Guoqiang 1990) is the speed of recent growth in personal and institutional consumption – vehicles, office machinery, eating and drinking, subsidies. Another familiar problem is that of energy supply and transport availability, to support the continued growth of the industrial economy. The problem of support also arises in agriculture, where 400,000 people annually have been leaving the land for urban jobs, and the province's cultivated land has been falling by around 27,000 hectares per year. Much is hoped for, in the governor's view, from foreign investment. Liaoning enjoys the largest share of foreign government loans in China, mainly made in large-scale construction projects (chemical plants, a reservoir, urban water supplies and so forth). There are also foreign joint ventures.

Geographically, Liaoning displays a characteristic which is becoming common in China – a prosperous industrial and urban core surrounded, at no great distance, by backward peripheries which require support and subsidy. In part this periphery comprises old-fashioned agricultural and pastoral areas, in part declining industrial zones and the weakest parts of the coalfield (Zhou Shulian and others 1990, 137–8). The tendency to depression of Liaoning's agriculture certainly owes something to the proximity of the surplus provinces of Jilin and Heilongjiang to the north. It also owes something, no doubt, to the long-established primacy of Liaoning in heavy industry, and by the same token also in the state industrial system.

The wider north-east has had an interesting development history under the People's Republic. In agriculture, the state has sponsored land clearance on a very big scale for decades in both Jilin and Heilongjiang, together with migration projects and a powerful emphasis on high-yielding maize crops and the functions of a commodity grain base. In industry, there has been large-scale development in Heilongjiang in both oil and coal. Both kinds of development have excited criticism from local committees and officials, as serving the objectives of the state rather than the best interests of the region. The local authorities would prefer to see less production for production's sake and more attention to the planning of whole communities; less primary production for supply to provinces further south, and more secondary and tertiary production to improve standards at home. In part they are no doubt influenced by the low prices paid for maize, oil and coal, by comparison with rapidly rising costs. Heilongjiang has complained particularly about losses in oil and coal production (Li Jie and Liu Xin 1990). Physical conditions meanwhile are difficult in Heilongjiang, with severe cold in winter, shortages of fuel, and house property often inadequate to give proper shelter in winter. From the state's point of view, however, it must be difficult to resist the opportunities for large-scale exploitation of new resources presented by the north-east.

Guangdong

A second province which displays exceptional local prosperity is Guangdong, the province of Guangzhou (Canton) and Hong Kong. The foundations of prosperity in Guangdong, however, are by no means the unexploited resource and state enterprise of Liaoning. In Guangdong, prosperity is based upon a flexible and many-sided collaboration between local families, businesses and even officials and their counterparts in Hong Kong. This collaboration has developed with remarkable speed since the agreement of 1984 for transfer of Hong Kong to Chinese rule in 1997.

Guangdong is a very important province. It probably contributes some 18 per cent of the positive figure in China's balance of payments (Ye Xuanping 1989, 15), against 5.4 per cent (60 millions) of the country's population, as well as contributing 24 per cent of foreign capital entering China (*CSY* 1990, 615). Hong Kong businesses now employ between 1 million and $1\frac{1}{2}$ million industrial workers in Guangdong, against only some 500,000 in Hong Kong itself. Guangzhou and Hong Kong are some 130 km apart, like Osaka and Nagoya or New York and Philadelphia. Figure 8.9 shows that in the Pearl River delta area, of which Guangzhou is the unquestioned centre, most of the big towns are on the west side of the estuary, but the Hong Kong–Guangzhou axis must develop in the course of time on the east side. There appear to be the materials here for a very powerful metropolitan centre, with a total population (including the delta) of around 13 million people at the present time.

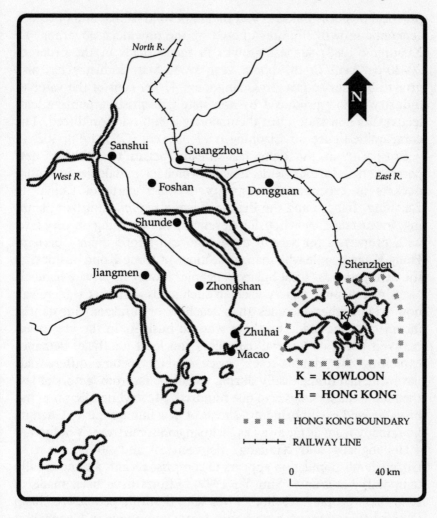

Figure 8.9 Hong Kong and the Pearl River delta, identifying places mentioned in the text

Guangdong – both the people and the provincial authorities – appears to have embraced the liberalization policies of the mid-1980s with enthusiasm; and the cutbacks of 1989 seem to have been accepted as a necessary, but by no means permanent, reaction. The delta is in a strong position in relation to official policy, which now favours capital and technology imports, stronger export earnings, practical business linkages within and beyond China, and of course economic growth. Figures quoted by the provincial governor (Ye Xuanping 1989) suggest growth in many fields of the order of 20–30 per cent in the boom year 1988. State industry has also grown, but not so fast. In Guangdong, 47 per cent of the value of industry is now produced by non-state enterprises (counting 'collectives' as non-state), but the majority is still state-produced. The comparable figure for Liaoning is 27 per cent (*CSY* 1990, 402).

Prosperity in the delta has already begun to establish new parameters in various fields, most of which have analogies in similar pockets of exceptional prosperity in other parts of China: in Liaoning, Jiangsu and the Beijing–Tianjin region. In urban planning, for instance, new housing developments are going up very fast, with properties for sale to those who can afford them – perhaps Hong Kong people who manage sections of Hong Kong business in the delta, perhaps local business people. Money in private hands is already reshaping society itself in such areas – market traders can now have higher incomes than teachers or officials. Substantial incomes can be made by those who do business in the grey areas between potential 'official' suppliers and local unofficial demand, exploiting the 'double-track' price system. Income differentials have widened dramatically during recent years, from factors of less than ten to factors closer to one hundred. Meanwhile migration has brought, and is still bringing, shoals of new immigrants to Guangzhou, many from as far away as Sichuan and some from even further – Heilongjiang and Xinjiang. Between 30 and 40 per cent of Guangzhou's population appears to comprise recent, and notionally temporary, migrants (Shu Yu 1989). Efforts have been made to send these people back home, but semi-destitute people are cheap to employ, and many have found means to survive in Guangzhou and the delta.

Township enterprises appear to be a key link in the system. Township enterprises represent cash incomes to both local peasants and local officials. They are also the usual means to subsistence for migrants in rural towns, in this representing central government policy. They produce cheap consumer outputs such as clothing, and if quality is high enough these outputs can also be exported. Hong Kong capital, or marketing opportunities, or linkages of various kinds, are usually involved. Local officials can usually avoid dog-fights with state enterprises about supplies or markets, and state enterprises may find the township enterprises useful customers for surplus coal or diesel. Mutual accommodations can often be arrived at in these fields (Woo 1991).

The agreement on Hong Kong has greatly strengthened Guangdong and the delta. Where central policy is the integration of Hong Kong with China, the central authorities cannot easily complain when the old Cantonese language and family ties between Hong Kong and the delta are revitalized. In any case, Guangdong received more than 2 million overseas Chinese visitors in 1988 (before the 4 June disaster of 1989) (*CSY* 1990, 620), and Hong Kong television is generally available in the delta. Maoist exclusiveness is now impossible. Further, if the central authorities wish to see continuing economic growth with overseas support where possible, they must wish to see Guangdong and the delta develop as they are now doing, and must be congratulated on the recognition that economic growth cannot be separated from social change which may be radical – though politics is a different matter.

The Special Economic Zones (SEZs) play a part in these developments, but increasingly it does not appear to be a central one. The most important of the SEZs, Shenzhen, lies just over the border from Hong Kong, and in appearance does not differ much from Hong Kong's new industrial suburbs. SEZs are supposed to attract foreign industrial investment, especially high-tech investment, in response to favourable conditions (of supplies, tax, export of profits and so forth) which can be negotiated with the officials. They have been somewhat disappointing in that most investment is from Hong Kong rather than from major foreign countries like the United States, and much less than was hoped is of high-tech character. Chinese policy, however, has had a hand in lowering the status of the SEZs. In 1986, Open Zones were declared in the Yangzi delta, the Pearl River delta and part of Fujian, and in 1988 Hainan (now a province) was added to these. Obviously the widening of special status tends to spread its benefits rather more thinly. In addition, it is now argued that a foreign firm may be able to get a better deal by negotiations in the delta with a local city or small town than in Shenzhen where the officials are less accommodating.

There is room to argue that throughout the south – the 'colonial' China of early imperial times – official control is broadly less unconditional than in the north, and the community less amenable to official direction and guidance. It has also been suggested that, conformably with this, the communist party's hold on the south has been weaker and less organic than elsewhere. To this 'south' the provinces of Guangdong and Fujian have been the extreme contributors. Both are the homelands of millions of overseas Chinese in south-east Asia and North America; both have long traditions of enterprise, foreign dealing, seafaring and piracy; both use Chinese dialects which are unintelligible to those of other communities and, of course, to northern Chinese. Both have taken on board the reforms of 1979 and developed them with exceptional energy. Guangdong particularly, with its close Cantonese ally Hong Kong, has made the reforms work with exceptional force, while keeping within Beijing's rubrics. In Guangdong the officials and the people are in alliance to make the best of the reforms, for the benefit of the province and its communities.

The mountains

To set against prosperous provinces, it is useful to look at the difficulties and limitations of regions of persistent poverty. The most important of these regions is the mountains.

Quite apart from the mountain ranges and plateaus of the central Asian half of China, a great part even of eastern China is occupied by mountain chains and high valleys. Approaching one-half of the whole area of eastern China lies above 1000 m elevation – most of the south, vast ranges in the centre such as those which isolate Sichuan, and high ranges in the north-east. Most mountain areas are poor, due to scarcity of farmland on the one hand and difficulties in transportation on the other. A commonly quoted traditional saw has it that the mountains are 'eight parts mountain, one part water, one part arable'; and grain may be very scarce. The 'eight parts mountain' may well be rich in timber or bamboo or minerals, but in these cases transport facilities are essential for the resources to be usuable, and of course as always with wild resources there are ecological needs to consider. Meanwhile the population continues to grow.

In many mountain areas, ecological problems have been inherited from previous centuries of exploitation, particularly the cutting of timber and bamboo and its replacement by scrub or grassland. A case in point is the Qinling region, whose 'capitalist' development under Ming and Qing has been outlined in chapter 3. During the past decades, population in this difficult region has continued to grow at a rate generally above 2 per cent per year, but grain output has grown less rapidly, generally at less than 2 per cent. A deficit has developed slowly but cumulatively, in spite of the opening of more land for cultivation – most of it, inevitably, of inferior quality. Alternative means of livelihood are limited, due to the stripping of the accessible forests in the earlier phase of settlement under Qing (Bao Jingcheng 1989). The Taihang escarpment, one of China's oldest habitats, has had a similar but parallel history of robbery of the environment in earlier centuries, including loss of timber as fuel for the smelting industries (Li Zhankui and others 1984). In the 13 provinces of south China, it is acknowledged that one-quarter of the total land area, around 67 million hectares, is occupied by low quality grassland. Most of this land must have been forested in earlier centuries.

Policy in these fields now has a number of features, mostly different from the emphasis on grain self-sufficiency of Mao's time. Labour migration is encouraged, with its corollary of remittances back home and (perhaps) the implication of later marriages. Commercial enterprises and the creation of new resources are also favoured, such as walnut or fruit orchards – stone fruits are often sold dried in China, and can be exported. On the southern grasslands, livestock farming is being promoted, though the scale of present development is tiny by comparison with the scale of the land area involved, or the scale of the transport and commercial networks necessary to bring livestock products to the market. The allocation of farmland under the household contract system seems certain to increase farm outputs in mountain conditions, and projects such as local reclamations of derelict land, the planting of orchards and enterprises such as the growing of seed potatoes (particularly suited to mountain conditions) and livestock or poultry rearing are also highly suited to household enterprise. Less so are road construction, local industries like paper-making and the maintenance of local hydro-power stations. Birth control is a particularly thorny issue in the mountain areas, especially since (unlike in Mao's time) households are now allowed to live outside the villages. Different Chinese authors take different views of the outlook of the populations, some deploring the survival of Maoist 'leftism', some their conservative and traditional thinking. Transport difficulties lead to very local outlooks among both people and officials, and poverty to low levels of education. 'Enterprise' is slow to make headway.

Enterprise is, however, a necessary part of the regeneration process, though not all of it. In Chinese discussions of the mountains in the 1950s, great emphasis was laid upon the capacity of the environments for variety of outputs and ways of life; and commentary has now returned to that position.

NINE

Retrospect and Prospect

Resource Scarcity for Ever?

'It was discovered in ancient times', wrote a distinguished sino-logist, 'that human demand must first be reduced in order to maintain peace in an economy of scarcity' (Yang 1952, 8). Yang was thinking about China under the traditional empire, but his comment is no less valid for the present. 'Economy of scarcity' is not the only way to describe contemporary China, but it has deep resonance. Resource is limited but potential population growth is not. During the past forty years technology has acted as a valuable lubricant for the friction and potential friction between resource and human demand, and it will continue to do so; but its potential is akin to that of resource – it is not finite, but it cannot be infinite.

Two groups of ideas seem to follow from Yang's insight. One relates to management. The 'peace' that is to be maintained must mean civil peace, and civil peace in China has to mean, in the historic empire and up to the present, official administration and management. Official management in turn must relate to a stable kind of society in which families and individuals accept their social and economic status, and social movement is limited. This is a historic ideal for Chinese society, and remains up to the present a more real practical ideal for the official class than would usually be declared. It still depends, perhaps, upon an intuitive notion of an economy of scarcity. Class struggle is supposed to be a thing of the past – and if of the future, that is not discussed.

The second set of ideas relates to the economy of scarcity itself. It has been expressed in recent years as 'resource scarcity for ever'. It has been shown in chapter 2 that although the Chinese subcontinent is not poor in resources, it inevitably appears so by the standards of the demands made upon it by the immense human community. The total population is likely to reach 1300 million (at least) by the year 2000, and there may then be latent unemployment in the countryside of the order of 300 million adult workers. The average of arable land availability will be of the order of 0.1 hectare per person (0.1 hectare equals 1000 square metres); of forest about the same, and of grassland around 2.5 hectares. Water supply may be below demand to the extent of about 10 per cent (Xiao Jiabao 1989). The prospects for energy output per person remain quite limited, as has been shown.

Apart from scarcities in all these important fields, due not so much to lack of resources or initiative in exploitation as to the scale of need, many of the problems of Chinese management cannot be separated from the demands of the huge population. Thus, state industry is stuffed with labour, because labour is always plentiful and people need jobs. Once migration of labour is tolerated, numbers migrating rise into the tens of millions. Improved personal hygiene involves vast quantities of water. City centres and public transport are crammed with people. The increase in the population continues, at a rate which the community cannot well afford. One reason why 'democracy' might perform worse than official rule in China is that 'democracy' would not be likely to maintain even the present flawed and less-than-minimum birth control policy, deeply unpopular as it is. At the present time, many other kinds of demand are also officially limited in China's economy of scarcity – for land, for raw materials, for foreign currency. Business opportunity is supposed to be available to all, but sometimes it too is limited by the officials. These are old principles in China, going back to recorded discussions of official management versus individual enterprise under the Han dynasty. People in China argue (as they do in the West) for different policies according to their different philosophical and social principles, and perhaps their vested interests.

The Modernization Possibilities

The Maoists persuaded themselves that China had little to learn from the West or even the Soviet Union. Western social systems were, it was thought, doomed to extinction, and in any case were not suited to Chinese conditions. But under the reformers since 1978, China has shown a marked interest in harnessing Western, Japanese and Hong Kong achievements, with the help of various administrative devices such as the Special Economic Zones, Open Zones and so forth. Most, if not all, Chinese provinces are said now to have promotional offices in Hong Kong, and many Guangdong units, even at county level, have the same. To its credit, the communist party accepts that Western technology cannot well be separated from other aspects of the Western economic environment, and it now displays a willingness to accept foreign individuals bringing in foreign management systems, for instance in factories run with foreign cooperation. It is true that in the Special Economic Zones (such as Shenzhen on the Hong Kong frontier) there is much less high-tech business and Western business than had been hoped, and more routine assembly work run by Hong Kong firms; but that ought to have been expected.

Modernization is the buzz-word in which these foreign adoptions find their legitimacy. But modernization is elusive. 'China is becoming modernised with amazing rapidity', wrote Carl Crow as long ago as 1933 (p.1). The fact is that modernization for China, as explained earlier, is an endless task which involves not only science on the land and efficiency in the factories, but the reconstruction of outlooks in a rapidly growing population numbering more than a billion, most of them very poor. Everything has to be done at once. And modernization is inevitably expensive, compared with the continuation of the old labour-intensive rural systems. At the same time, voices are now raised against the limitations of the old rural Chinese system towards the population – 'the unstoppable stream of low intelligence', according to one writer, who points out the weaknesses of marriages among close relatives (common in the countryside), high drop-out rates from school, continuing difficulties with superstition, congenital hereditary disease, poor nutrition of infants and illiteracy and semi-literacy (Gao Ping 1990). All this, it might be said, arises from low and static levels of modernization in the countryside, and technical and economic modernization are poorly placed to begin in these communities. In this context, the communist party's movement of 1985 towards commercialization of the rural production systems, with its concomitants of the shake-out of rural labour and the stimulation of the township enterprise sector, was a powerful movement in the direction of modernization, though at a potentially fearful cost in various ways – in loss of farm production, in hazard to the social stability whose value to the official class has already been pointed out, in urbanization and the growth of the great cities as well as small towns, in greatly increased pollution of all kinds. In twenty years' time, if policy remains stable, China will have many more urban workers (though the rural work-force may still be above need), many more factories of which more will be capable of supplying overseas markets, many more people doing tertiary work both urban and rural (a great and much-needed improvement to the quality of life) and, in the countryside, it is hoped, a class of skilled and specialist farmers supplying the cities and the state, and using farm technology which is much closer to Japanese standards. Modernization in these conditions will inevitably have more in common with the rest of the Third World than with the tidy technical and managerial changes to which the communist party would like to limit it.

Other and even more far-reaching changes may be in the pipeline. People are beginning to say that the present generation of young adults is different from all its predecessors, and that the call for democracy of 1989 had profound resonance as well as simple noise.

References

Academy of Agricultural Sciences (1990) A study on the balance of China's regional grain production and demand. *Nongye jingji wenti (Problems of Agricultural Economics)*, 1990/6, 46–9 (translated in JPRS *China Report*, 25 October 1990, 57–62).

An Zhiwen (1985) A year in which marked progress was made in the reform of the economic structure in cities. *People's Daily*, 13 May (translated in FBIS *Daily Report*, 22 May 1985, K6–K11).

Atlas of Natural Geography (1984) *Zhongguo ziran dili tuji (Atlas of Natural Geography of China)*. Beijing: North Western Teachers' University.

Aubert, C. (1990) The agricultural crisis in China at the end of the 1980s. In J. Delman, C. S. Østergaard and F. Christiansen (eds), *Remaking Peasant China*. Aarhus: Aarhus University Press, 16–37.

Bao Jingcheng (1989) Man–land relationships in the Qinling foothills. *Jingji dili (Economic Geography)*, 8 (3), 209–11.

Beijing Atlas (1984) *Zhonghua renmin gongheguo dituji (Atlas of the People's Republic of China)*. Beijing: Cartographic Press.

Beijing broadcast (1984) Obstacles to rural commodity production (translated in FBIS *Daily Report*, 3 April 1984, K10–K14).

Cannon, T. (1990) Regions: spatial inequality and regional policy. In T. Cannon and A. Jenkins (eds), *The Geography of Contemporary China*. London: Routledge.

Chen Dunyi and others (1983) *Zhongguo jingji dili (Economic Geography of China)*. Beijing: Zhanwang.

Chen Jian (1985) An effective way to increase employment. *Jingji yanjiu cankao ziliao (Reference Material on Economic Research)*, 114 (translated in *Chinese Economic Studies*, Fall 1987, 45).

Chen Yaobang and others (1989) *Zhongguo xiangzhen qiye nian jian (Yearbook of Township Enterprises)*. Beijing: Agricultural Press.

Chinese Communist Party (1984) *CCP Central Committee's Circular on Rural Work in 1984*, 1 January (translated in FBIS *Daily Report*, 13 June 1984, K1–K11).

Chinese Communist Party (1985) Ten policies of the CPC Central Committee and the State Council for further invigorating the rural economy, 1 January (translated in FBIS *Daily Report*, 25 March 1985, K1–K9).

Chinese Economic Studies (1987–8) The private economy, 1 and 2 (Fall 1987 and Winter 1987–8 issues, edited by Stanley Rosen, containing collected articles from the Chinese press).

Cressey, G. (1934) *China's Geographic Foundations*. New York: McGraw Hill.

Crow, C. (1933) *Handbook for China*. Hong Kong: Kelly and Walsh (reprinted by Oxford University Press, 1984).

CSY (1990) *China Statistical Yearbook 1990* (translated). Chicago: University of Illinois. Earlier editions from Economic Information Agency, Hong Kong (these works continue the series called *Statistical Yearbook of China*; both are translations of the current *Tongji nianjian* volumes).

Deane, P. (1979) *The First Industrial Revolution*, 2nd edn. Cambridge: Cambridge University Press.

Deng Yiming (1991) Predicaments in the commodity grain bases, and ways out. *Nongye jingji wenti (Problems of Agricultural Economy)*, 1991/1, 21–5.

Development Research Institute, Production and Enterprise Unit (1986) Development of rural non-agricultural production. *Jingji yanjiu*, 1986/8, 9–24.

Durand, J. D. (1960) The population statistics of China, AD 2–1953. *Population Studies*, 13 (3), 209–55.

Dwyer, D. (1986) Urban housing and planning in China. *Transactions, Institute of British Geographers*, NS 11 (4), 479–89.

Elvin, M. (1972) The high level equilibrium trap: the causes of the decline of invention in the traditional textile industries. In W. E. Willmott (ed.), *Economic Organization in Chinese Society*. Stanford, CA: Stanford University Press.

Fang Dahao and others (1990) Exploration of some issues connected with the enterprise contract system. *Gaige (Reform)*, 1990/1, 147–52 (translated in JPRS *China Report* 13 April 1990, 19–21).

Fang Xing (1979) The rise and decline of 'spouts of capitalism' in Shaanxi under Qing. *Jingji Yanjiu (Economic Research)*, 1979/12, 59–67.

Feuchtwang, S., Hussain, A. and Pairault, T. (1988) *Transforming China's Economy in the Eighties*, two volumes. Boulder, CO: Westview Press;

London: Zed Books.

Gao Ping (1990) A study and reflexion on the quality of China's population. *Ban Yue Tan (Fortnightly)*, 10 May (translated in JPRS *China Report*, 6 August 1990, 55–8).

Geographical Research Institute, Chinese Academy of Sciences (1980) *Zhongguo nongye dili zong lun (Agricultural Geography of China)*. Beijing: Science Press.

Han Shu (The History of Han) By Pan Gu of the first century AD, and relating mainly to the two preceding centuries. The Shi Huo sections, on economic life, are translated by N. L. Swann (1950) *Food and Money in Ancient China*. Princeton, NJ: Princeton University Press.

Ho Ping-Ti (1959) *Studies in the Population of China*. Cambridge, MA: Harvard University Press.

Hodder, R. N. W. (1990) China's industry – horizontal linkages in Shanghai. *Transactions, Institute of British Geographers*, NS 15 (4), 487–503.

Hou Zhemin and Zhang Qiguang (1989) Peasants – government – agricultural problems. *Jingji Zhoubao (Economics Weekly)*, 21 May, 2 (translated in JPRS *China Report*, 19 July 1989, 37–40).

Hou Xueyu (1979) *Zhonghua renmin gongheguo zhibei ditu (Map of Vegetation Cover of the PRC)*, 1 : 4 million. Beijing: Cartographic Publishing House.

Howe, C. (1968) The supply and administration of urban housing in mainland China – the case of Shanghai. *China Quarterly*, 33, 73–97.

Hu Angang (1989) *Renkou yu fazhan (Population and Development)*. Hangzhou: Zhejiang People's Publishing House.

Huang Daoxia (1987) Is there polarisation in the rural areas? *Ban Yue Tan (Fortnightly)*, 25 November, 20–3 (translated in JPRS *China Report*, 8 December 1987, 26–8).

Hughes, T. J. and Luard, D. E. T. (1961) *The Economic Development of Communist China*, 1949–60. London: RIIA.

Jin Yong (1987) The 'Nihongling' passenger bus – when can it run without hitches? *Chinese Economic Studies*, 21 (2), 50–2.

Jowett, J. A. (1989) Mainland China: a national one-child program does not exist, parts 1 and 2. *Issues and Studies*, 25 (9), 48–67; 25 (10), 71–97.

Kirkby, R. J. R. (1985) *Urbanisation in China: Town and Country in a Developing Economy, 1949–2000 AD*. London: Croom Helm.

Ko Din and Liu, S. (1990) Wheels of Misfortune. *China Trade Report*, March, 6–7.

Küchler, J. (1990) On the establishment of a poverty-oriented rural development policy in China. In J. Delman, C. S. Østergaard and F. Christiansen (eds), *Remaking Peasant China*. Aarhus: Aarhus University Press, 124–38.

Lardy, N. R. (1978) *Economic Growth and Distribution in China*. Cambridge: Cambridge University Press.

Leeming, F. (1985a) *Rural China Today*. Harlow: Longman.

Leeming, F. (1985b) Chinese industry – management systems and regional structures. *Transactions, Institute of British Geographers*, NS 10 (4), 413–26.

Leeming, F. and Powell, S. (1990) Rural China: old problems and new solutions. In T. Cannon and A. Jenkins (eds), *The Geography of Contemporary China*. London: Routledge, 133–67.

Lei Xilu (1990) Investigation of questions of migration of surplus labour in China. *Nongye jingji wenti (Problems of Agricultural Economics)*, 1990/4, 30–2.

Leung, C. K. (1980) China: railway patterns and national goals. University of Chicago Department of Geography Research Paper no. 195.

Li Guoqiang (1990) Interview with Li Changchun, governor of Liaoning province. *Kuang chiao ching (Wide Angle)*, 208, 46–51 (translated in JPRS *China Report*, 7 May 1990, 34–8).

Li Jie and Liu Xin (1990) Causes and countermeasures for the decline in economic effectiveness of Heilongjiang's industry. *Heilongjiang ribao (Heilongjiang Daily)*, 16 September (translated in JPRS *China Report*, 30 November 1990, 27–8).

Li Jinchang (1988) *Woguo ziyuan yu huanjing (China's Resources and Environments)*. Beijing: Xinhua Books.

Li Shusheng (1989) The problem of capital for the procurement of farm products. *Caimo jingji (Finance and Trade Economics)*, 1, 33–9 (translated in JPRS *China Report*, 19 July 1989, 28–36).

Li Tiezheng (1990) Cotton, where are you? *Jingji ribao (Economic Daily)*, 23 July (translated in JPRS *China Report*, 10 October 1990, 64–5).

Li Yongzeng (1989) Two-way view of China's petroleum industry. *Liaowang Overseas*, 13 March, 8–11 (translated in FBIS *Daily Report*, 11 April 1989, 34–6).

Li Zhankui and others (1984) Bringing people's initiative into full play to speed up the economic construction of the Taihang mountainous areas. *Nongye jingji wenti (Problems of Agricultural Economics)*, 1984/4, 3–9.

Li Zhisheng (1989) The energy crisis – present and future. *Guanli shijie (Management World)*, 1989/1, 120–9 (reprinted in *Industry*, Chinese People's University Materials Centre, 1989/3, 149–58).

Lippitt, V. D. (1978) The development of underdevelopment in China. *Modern China*, 4, 3.

Lipton, M. (1977) *Why Poor People Stay Poor. A Study of Urban Bias in World Development*. London: Temple Smith.

Liu Xiejang (1989) Where is coal's stimulus to come from? *People's Daily*, 5 March.

Liu Zifu and Wang Man (1989) Problems in grain procurement analysed. *Nongmin ribao (Peasant's Daily)*, 31 March (translated in FBIS *Daily Report*, 17 April 1989, 58–61).

Lockett, M. (1988) The urban collective economy. In S. Feuchtwang, A. Hussain and T. Pairault (eds), *Transforming China's Economy in the Eighties*, volume 2. Boulder, CO: Westview Press; London: Zed Books, 118–37.

Loewe, M. (1966) *Imperial China*. London: Allen and Unwin.

Lu Longwen and Chi Tingxi (1983) Gao Yangwen speaks about prospects for China's coal industry. *Shijie jingji daobao*, 21 February, 2 (translated in FBIS *Daily Report*, 11 March 1983, K7–K10).

Lu Xueyi and Zhang Houyi (1990) Peasant diversification – problems, remedies. *Nongye jingji wenti (Problems of Agricultural Economics)*, 1990/1, 16–21 (translated in JPRS *China Report*, 29 May 1990, 59–67).

Luo Yuanming (1990) Reforming the management system for state-owned assets. *Jingli guanli (Economic Management)*, 1990/4, 28–31.

Ma Rong (1990) Role of small cities and towns in modernisation. *Zhongguo shehui kexne (Chinese Social Sciences)*, 1990/4, 131–46 (translated in JPRS *China Report*, 7 September 1990, 26–37).

Ma Xiangyong and others (1984) Some crucial questions in farm production in the Lake Tai basin. *Dili xuebao (Acta geographica sinica)*, 39, 1.

Mao Zedong (ed.) (1956) *Zhongguo nongcun de shehuizhuyi gaochao*. Beijing (translated as *Socialist Upsurge in China's Countryside*. Beijing: Foreign Language Press).

Middle School Atlas (1978) *Zhongguo dituce (Map Book of China)*. Shanghai: Cartographic Press.

Ning Ke (1979) Agricultural production in the Han dynasty. *Guangming Daily*, 4 October, 4.

Nishijima Sadao (1984) The formation of the early Chinese cotton industry. In L. Grove and C. Daniels (eds), *State and Society in China: Japanese Perspectives on Ming-Qing Social and Economic History*. Tokyo: University of Tokyo, 1–17.

Nongye jingji wenti (1986) Editorial: Gradually implement the strategic transfer of agricultural labour. *Nongye jingji wenti (Problems of Agricultural Economics)*, 1986/10, 3.

Oi, J. C. (1986) Peasant grain marketing and state procurement – China's grain contracting system. *China Quarterly*, 106, 272–90.

Perkins, D. H. (1969) *Agricultural Development in China, 1368–1968*. Edinburgh: Edinburgh University Press.

Qi Jingfa and others (1990) Discussion of peasant burdens. *Jingji ribao*

(Economic Daily), 3 February (translated in JPRS *China Report*, 23 March 1990, 72–4).

Shanghai Atlas (1984) *Zhonghua renmin gongheguo dituji (Atlas of the People's Republic of China)*. Shanghai: Cartographic Publishing House.

Sheng Chengyu and others (1986) *Zhongguo qihou zonglun (General Review of Chinese Climates)*. Beijing: Science Press.

Shi Wei (1990) Research on the relationship between vegetable production and vegetable subsidies in Beijing. *Zhongguo nongcun jingji (China's Rural Economy)*, 1990/6, 55–8 (translated in JPRS *China Report*, 25 October 1990, 62–5).

Shu Yu (1989) One in twenty of China's population is on the move. *Renmin ribao (People's Daily)*, 26 February.

Shue, V. (1990) Emerging state–society relations in rural China. In J. Delman, C. S. Østergaard and F. Christiansen (eds), *Remaking Peasant China*. Aarhus: Aarhus University Press, 60–80.

Sit, V. F. S. (ed.) (1985) *Chinese Cities*. Hong Kong: Oxford University Press.

Smil, Vaclav (1984) *The Bad Earth*. Armonk, NY: M. E. Sharp; London: Zed Books.

Song Qing (1990) Rebound effect of the grain–cotton price ratio mechanism. *Jingji ribao (Economic Daily)*, 17 August (translated in JPRS *China Report*, 10 October 1990, 61–3).

Statistical Materials 1949–1984 (1985) *Zhongguo gongye de fazhan – tongji ziliao 1949–84 (Development of Chinese Industry – Statistical Materials, 1949–84)*. Beijing: State Statistical Bureau, Industry and Transport Statistics Unit

Sun Hao (1986) Unhealthy trends in the development of township collieries must be stopped. *Jingli guanli (Economic Management)*, 1986/2, 24–5 (translated in FBIS *Daily Report*, 30 April 1986, K11–K15).

Sun Jingzhi (ed.) (1988) *The Economic Geography of China*. Oxford: Oxford University Press.

Sun Zhonghua (1990) A glimpse of grain production from 1984 to 1988 – a survey of the grain output of 13,000 peasant households in 155 villages (a survey authorised by the Rural Development Research Centre, State Council). *Zhongguo nongcun jingji (China's Rural Economy)*, 1990/3, 16–24 (translated in JPRS *China Report*, 26 June 1990, 73–80).

SYC (1986, and earlier editions) *Statistical Yearbook of China*, State Statistical Bureau, PRC. Hong Kong: Economic Information Agency; Oxford: Oxford University Press (these works are forerunners of the *China Statistical Yearbooks*; both are translations of current *Tongji nianjian* volumes).

Tang Zongkun (1987) Supply and marketing. In G. Tidrick and Chen Jiyuan (eds), *China's Industrial Reform*. New York: World Bank; Oxford: Oxford University Press, 210–36.

Ten Great Years (1960) Translated from *Weida de shi nian*, 1959, compiled by State Statistical Bureau.

Teng, S.-Y. and Fairbank, J. D. (1963) *China's Response to the West – a Documentary Survey, 1839–1923*. New York: Atheneum.

Tidrick, G. and Chen Jiyuan (eds) (1987) *China's Industrial Reform*. New York: World Bank; Oxford: Oxford University Press.

Urban Social and Economic Survey Organisation, State Statistical Bureau (1990) *China, Forty Years of Urban Development*. Beijing: China Statistical Information and Consultancy Services Centre.

Wang Mengkui (1991) Interview with Wang (Deputy Director of the State Council's Research Office). *Liaoning Overseas*, 1991/5 (translated in FBIS *Daily Report*, 19 February 1991, 32–5).

Wang Ping and Song Qing (1990) On the necessity and feasibility of smoothing out the price priorities between pigs and grain. *Jingji yanjiu (Economic Research)*, 1990/5, 78–80 (translated in JPRS *China Report*, 28 September 1990, 47–50).

Wang Xianjin (1989) The current situation and trends in the development of farmland in China, and measures to be adopted. *Renmin ribao (People's Daily)*, 24 July, 6 (translated in JPRS *China Report*, 8 August 1989, 46–9).

Wang Xiaokun (1989) Interview with Wang Xianjin, Director of the State Land Administration Bureau. *People's Daily*, 13 April.

Woo, E. (1991) Unpublished research for the PhD degree, School of Geography, University of Leeds.

Wu Fumi and Li Zhiyong (1990) Entering the 1990s with confidence. *Liaowang (Outlook)*, 1990/9, 4–6 translated in JPRS *China Report*, 11 July 1990, 62–5).

Wu Huailian (1989) The tide of peasants leaving the land in the eighties. *Renkou xuekan (Population Journal)*, 1989/5, 41–9.

Xian broadcast (1985) Report on Li Peng's visit to Shenmu, 15 November (translated in FBIS *Daily Report*, 25 November 1985, K5).

Xiao Jiabao (1989) Four major crises which China will face, and countermeasures. *Liaowang, Overseas Edition*, 6 March (translated in FBIS *Daily Report*, 19 April 1989, 28–34).

Xiao Zhenghong (1988) The rise and decline of agriculture in southern Shaanxi under Qing. *Zhongguo nong shi (Chinese Agricultural History)*, 1988/4, 69–84.

Yan Xiaofeng and others (1990) Survey of changes in social structure since reform. *Zhongguo shehui kexue (Chinese Social Sciences)*, 1990/4, 121–30 (translated in JPRS *China Report*, 7 September 1990, 37–45).

Yang, L. S. (1952) *Money and Credit in China*. (Cambridge, MA: Harvard University Press.).

Ye Xuanping (1989) Government work report by governor Ye Xuanping, Provincial People's Congress, 2 March (translated in JPRS *China Report*, 12 March 1989, 13–29).

Yu Chengsheng (1985) The Hanjiang River and the north–south water transfer project. *Dili yanjiu (Geographical Research)*, 4 (2), 89–94 (translated in *Chinese Geography and Environment*, Spring 1988, 31–46).

Zhang Du and Liu Pingchun (1988) Money has been spent and prices have risen; why are there fewer vegetables? *Hunan ribao (Hunan Daily)*, 2 November (translated in JPRS *China Report*, 23 February 1989, 44–5).

Zhang Ping (1990) An analysis of the effects of price subsidies on the distribution of residents' incomes in China. *Jingji yanjiu (Economic Research)*, 1990/4, 36–43 (translated in JPRS *China Report*, 6 August 1990, 26–33).

Zhao Songqiao (1986) *Physical Geography of China*. Beijing: Science Press; New York: John Wiley.

Zhong Xiaopo (1989) Schemes to regulate the distribution of vegetable outputs in China. *Jingji dili (Economic Geography)*, 9 (2), 97–100.

Zhou Shulian and others (1990) *Zhongguo diqu chanye zhengei yanjiu (Studies of Regional Production Systems in China)*. Beijing: China Economic Press.

Zhu Shuofu (1989) Reasons for China's scarcity of coal and means to resolve it. *Jingji guanli (Economic Management)*, 1989/8, 21–2.

Guide to Further Reading

General Surveys

Cannon, Terry and Jenkins, Alan (eds), 1990. *The Geography of Contemporary China*, Routledge, London and New York. Geographical focus.

Feuchtwang, Stephan, Hussain, Athar and Pairault, Thierry (eds), 1988. *Transforming China's Economy in the Eighties*, Zed Books, London and Westview Press, Boulder, Col. Primarily economic and sociological focus.

Goodman, David (ed.), 1989. *China's Regional Development*. Routledge and RIIA, London and New York. Varied focus: economic, regional, political.

Dwyer, Denis (ed.), forthcoming. *China: The Next Decades*. Geographical focus.

Smith, Christopher J., 1991. *China. People and Places in the Land of One Billion*. Geographical focus, with varied (social, political) background material.

Hook, Brian (ed.), 1991. *The Cambridge Encyclopedia of China*, 2nd ed. Cambridge University Press, Cambridge. Wide-ranging and authoritative.

Individual Topics

Zhao Songqiao, 1986. *Physical Geography of China*. Science Press, Beijing and John Wiley, New York. Excellent review of China's environments, handsomely produced and illustrated.

Smil, Vaclav, 1984. *The Bad Earth*. Sharp, Armonk and Zed Books, London. Pollution and resource scarcity in China.

Blunden, Caroline and Elvin, Mark, 1983. *Cultural Atlas of China*. Phaidon, Oxford. History and rationale of Chinese culture and society, both traditional and modern.

Hinton, William, 1991. *The Privatisation of China*. Monthly Review Press, New York, 1990; Earthscan, London. Recent developments in social and economic organization.

Oi, Jean C., 1989. *State and Peasant in Contemporary China*. University of California Press, Berkeley and Oxford. In-depth analysis of the working of the Chinese countryside, especially its management.

Delman, Jorgen, Ostergaard, Clemens and Christiansen, Flemming (eds), 1990. *Remaking Peasant China*. Aarhus University Press, Aarhus. Several original and important contributions.

Fei Xiaotung, 1989. *Rural Development in China – Prospect and Retrospect*. Chicago University Press, Chicago and London. Essays on villages and small towns, by China's best-known sociologist.

Tidrick, Gene and Chen Jiyuan (eds), 1987. *China's Industrial Reform*. World Bank and Oxford University Press, New York and London. Analysis of problems in Chinese industry based mainly on a series of field studies.

Kirkby, Richard, 1985. *Urbanization in China*. Croom Helm, London. Chinese urbanization in practice and theory.

Sit, Victor F. S., 1985. *Chinese Cities*. Oxford University Press, Hong Kong. Studies of individual cities.

Whyte, Martin King and Parish, William L., 1984. *Urban Life in Contemporary China*. University of Chicago Press, Chicago. Day-to-day and year-to-year experience in urban China.

Materials from China

Luo Hanxian, 1985. *Economic Changes in Rural China*. New World Press, Beijing. Optimistic view of Chinese rural experience since 1949.

Sun Jingzhi, 1988. *The Economic Geography of China*, Oxford University Press, Hong Kong. Translated from *Zhongguo jingji dili zong lun*, Beijing 1984. Standard Chinese economic geography; high on simple description, low on continuing problems.

Xue Muqiao, 1981. *China's Socialist Economy*. Foreign Languages Press, Beijing. Still the best review of 'reform' policy.

State Statistical Bureau, PRC. Annual. *China Statistical Yearbook*. Chinese Statistical Information and Consultancy Service and University of Illinois, Beijing and Chicago. The translated version of the standard Chinese statistical annual, almost identical with the Chinese-language version.

Map of China – *Zhonghua renmin gongheguo ditu* (*Map of the People's Republic of China*), 1981. Ditu Chubanshe, Beijing. Handsome one-sheet map on 1/6 million scale; excellent relief, names in *pinyin* romanization.

Index